MEMOIRS
Of A
PHOENIX

Based On The True Story
On Daniel Anthony Moreno

DANIEL MORENO

Memoirs Of A Phoenix: *Based On The True Story On Daniel Anthony Moreno*

ISBN: 978-1-948382-53-3 paperback

Jones Media Publishing
10645 N. Tatum Blvd. Ste. 200-166
Phoenix, AZ 85028
www.JonesMediaPublishing.com

Printed in the United States of America

Praise For
Memoirs of a Phoenix

"Memoirs of a Phoenix by Daniel Moreno – this is a difficult book to read. Throughout the book you can see a victim being overcome by life. From drugs, relationships, and even a life threatening brain injury. You just wonder when this will end. Then in the middle of it all you see God's love and His hand of grace, as He leads Daniel through the restoration and instills in him a new joy in life."

— Michael Williams, pastor

Dedication

I would like to dedicate this book with 22 chapters to the 22 veterans that commit suicide every day. I was very close to being one of them but simply chose for my story not to end there. Also to my mom and God. She made me the man I am today but God put me through a process. "The gem cannot be polished without friction nor a man without trials" (Seneca).

"At times the world may seem an unfortunate and sinister place. But I believe there is much more good in it than bad. All you have to do is look hard enough. And what seems to be a series of unfortunate events may in fact be the first steps to a journey" (Lemony Snicket). Google #sapper_ninja to see most of the content in this book.

Table of Contents

Foreword

I was honored to serve with Daniel Moreno and all the other exceptional men and women in the 44th Engineer Battalion – The Broken-heart Battalion. We served together as Soldiers during a time of many historic events and I'm proud of each of them for their selfless service to our country.

The 44th Engineer Battalion was stationed at Camp Howze, South Korea during the US invasion of Iraq in the early stages of Operation Iraqi Freedom in 2003. During the rise of the Iraqi insurgency in 2004, our Brigade Combat Team – The 2nd Brigade Combat Team (BCT), 2nd Infantry Division, was deployed from South Korea to Ramadi, Iraq in the Sunni Triangle. Ramadi is the capital and largest city of Iraq's western Anbar province. It is located in the upper Euphrates river valley, situated mostly on the Southern banks of the river, about 70 miles west of Baghdad and it was the hotbed of the insurgency.

The deployment of our BCT from South Korea to Ramadi, Iraq was historic because US units had previously never deployed from South Korea into Combat in another country. Our Army BCT was attached to the 1st Marine Division and a Marine battalion (2nd Battalion, 4th Marine Regiment) was attached to our BCT to increase our combat power. We arrived in September 2004 at the height of the insurgency when the total US Military Deaths in Iraq reached 1000. We were immediately embroiled in combat besieged by daily mortar and rocket attacks against our base camp and IEDs and ambushes during our combat missions clearing routes in and around the city, searching for enemy weapons caches, and providing security for the population against the insurgency. By November 2004 the fighting led to the deadliest month to date for US forces with over 137 deaths.

The Battalion's Soldiers fought bravely and performed exceptionally well as did all the Soldiers in our Brigade Combat Team. Their efforts had the intended effect as security slowly increased. By JAN 2005, Iraq was able to conduct historic elections for their National Assembly, the first free national elections since the fall of Saddam Hussein. Our Soldiers helped establish the voting stations and provided security to ensure voters were safe from insurgents as they cast their votes in this important election. Security continued to increase throughout our one year tour of duty. By the time we departed in August 2005, mortar and rocket attacks against our base camp had decreased from the daily attacks we first encountered to the point we'd go for weeks between attacks and engagements with the insurgents became much less frequent during our daily security operations.

The success of our operations provided a much more stable and secure environment for the Iraqi citizens of Ramadi but came at a significant cost. Our battalion lost 9 Soldiers killed in action while fighting in Ramadi – SPC Robert O. Unruh, PFC Aaron J. Rusin, SSG Omer T. Hawkins, PFC Mark A. Barbret, PFC Andrew M. Ward, SSG Arthur E. Williams, SGT Bennie J. Washington, SSG Andrew T. Bossert, and PFC Michael W. Franklin in addition to many others who were wounded. Additionally CPL Piotr Szczypka was killed in a tragic assault soon after our redeployment to Fort Carson, Colorado. Each of these young men were exceptional, they were our brothers and their loss continues to haunt and hurt all of us who served with them. Their lives were cut short and their dreams and future aspirations unfulfilled.

Each of the 627 Soldiers who served in the Broken-heart Battalion has their own unique story. A story that includes how and where they were raised, the defining moments in their lives, why they joined the Army, how they dealt with the combat they faced, and how the experience affected and changed them forever. The story also includes family, parents and grandparents, love relationships, children, friends, failures, challenges, and triumphs.

This book contains one such story. It's the story of an incredible young man, Daniel Moreno, told in his own words. It's the story of a son and grandson who was raised, nurtured, and loved by his mother, grandmother, and step-father. It's a story of a brother who loves his siblings. It's the story of a Soldier who served faithfully, fought for our country, and cared for his brother and sister Soldiers. It's a story that includes the loss of the 10 Broken-heart Soldiers he served with who died in Iraq and Ft. Carson. The story of their loss became a part of the story of all of us who served with them.

But most of all it's an amazing story of a survivor who continues to persevere and overcome all challenges life throws at him. And with each challenge overcome, he becomes stronger and more resilient like the mythical Phoenix rising above the ashes. He overcame a broken home with a father in and out of prison, expulsion from school, bullying and bad associations with thuggish friends involved in petty crimes and drugs to become a Soldier, a combat engineer, in the US Army. He overcame the fear, anxiety, apprehension, and uncertainty associated with his early combat experience to become an outstanding, confident, competent Soldier who excelled in Iraq. He overcame the grief and guilt of losing brother Soldiers in combat and PTSD and continued to serve.

He attended and graduated from some of the Army's most demanding schools – Airborne School where he gutted out difficult runs and airborne operations despite an injured ankle, The Primary Leadership Development School, and Sapper School. He competed in the Army's elite Sapper Leader Competition. He became a Sergeant and a leader and was preparing for Ranger School when an Arteriovenous Malformation the size of a large plum was found in the right temporal lobe of his brain. This subsequently led to a stroke and removal of 30% of his brain. Once again he overcame this seemingly insurmountable challenge staying positive through strenuous rehabilitation to adjust to the loss of half his peripheral vision, reduced cognitive abilities with long and short term memory loss, impaired balance, and coordination. While recovering, he also had to overcome the painful breakup of his marriage and separation from his wife and children. These very difficult challenges led to thoughts of suicide. His newfound faith, the love and support of his mother, and his incredibly strong resilience and positive attitude allowed him to continue his recovery. Ultimately his recovery was so successful he was able to compete in incredibly challenging Ninja Warrior Competitions. He continues to live and enjoy life to the fullest today.

He also continues to care about those who served our nation in uniform and has devoted this book and its 22 Chapters to the veterans who commit suicide each day. I believe Daniel's story provides insights helpful to understanding this crisis. I know first-hand how Veteran's experiences can affect them. While all the Soldiers in the 44th Engineer Battalion left Ramadi, Iraq in August 2005, Ramadi never left us. Our time there, those we served with, the emotions we felt, and the experiences that changed us are never far from our mind and sometimes can appear overwhelming.

I thank Daniel for writing this book and sharing his experiences and thoughts in such an unvarnished way. Reading this book was very meaningful for me. I am proud to have served with Daniel in the 44th Engineer Battalion, I am honored to have remained in contact with him over the past 17 years, and I continue to be inspired by him.

TOMMY R. MIZE
COL(R), United States Army

Prelude

F*ear*. Primal, guttural, *fear*. When I first stepped foot on Iraqi soil, the overwhelming feeling that overtook and lurched within the pit of my stomach was that of fear. Fear clouded over the workings of my mind and cast a long and dark shadow over every little thing. Not only did I recognize fear bubbling inside me and in the goosebumps that crawled all over my skin as we approached Iraq, I saw it in the blank faces of the soldiers around me as well. Although no one said a word out of the ordinary, behind the reserved introductions that we all shared, I could sense that my fellow brothers-in-arms were almost as afraid as I was. I suppose in that regard; especially fear works in mysterious ways. One day, you're packing up and saying goodbye to your family, all bright-faced and eager to serve your country; and the next, you're approaching Iraq, and suddenly, out of thin air, fear crawls up and takes a strong hold of you. If you've ever watched movies about soldiers being stationed in Iraq or Afghanistan, you might be expecting us to burst with valiance and bravery; and while that is true to a certain extent, it definitely isn't the entire picture. As a soldier serving my country, Iraq caught me off guard.

I was stationed in Ramadi, Iraq, in September 2004, and I was deployed in that area for a period of twelve months. This was my first and only deployment to the area. For the first two weeks of that time, I was completely gripped by apprehension. I could feel it while conducting our missions. I know this isn't what you expect to read from an autobiography written by a soldier, but it's the truth. I mean, I was human after all. As I write this and look back on my first few days in Iraq, I remember a lot, especially the terror that I felt. Out of all emotions, fear is perhaps the most dangerous. Not only does it catch you at a time when you're blissfully unaware of its existence, but it also does so with such power that you often find yourself wondering if there ever was a time when you weren't so afraid.

Fear seeps and shadows over everything, and in many ways, acts like a slow poison that distorts and decays your rational mind. In my case, however, I lucked out. As time continued to steer itself forwards and I got used to the Iraqi soil, I found that every day that I woke up, I woke up feeling a lot more rejuvenated than I was the night before. It was almost as if I'd just gained a new sense of clarity, and as fear loosened its grip on me, I got accustomed to the day-to-day routine of my life as a soldier stationed in foreign a hostile land. I began to realize the weight of my responsibility and acted accordingly.

However, after the first two weeks, I was immediately thrust into on-field action on our route clearance mission. In those days, smoke and fire seemed to follow me everywhere that I went. There were bombings and explosions everywhere around me, and as time went on, I got used to hearing their fierce sound and responding accordingly. As I followed my sergeant's orders and carried out my duty as a soldier as best as I could, my nostrils would fill up with the smell of smoke and dust during the time that I spent in Iraq. My ears got accustomed to the constant sound of bombing and gunfire that went off. Truth be told, Iraq took a toll on me. We were being shot at constantly, and RPGs were being shot at us with an alarming frequency. One of the biggest lessons that I learned during my time as a soldier was to always be on the lookout head on a swivel, and looking back at it, I can definitely say that Iraq had a major part to play in teaching me that lesson. Being on Iraqi soil, we were in dangerous territory, and so the arrival of imminent threats wasn't a matter of *how*; it was a matter of *when*. Having said that, although I no longer felt the fear that I did when I was first stationed, there were still a lot of incidents that I had trouble dealing with the consequences of.

Two of these incidents, which I will probably never forget as long as I live, was when a fifty-gallon drum blew up across the street from me where one of my sergeants wanted to position me. It loosened me up. Also the other that happened a few days later when we were pulling security from the safety of the M113. I was completely frustrated by our slow-paced retaliation tactics. Anger pulsed through my veins, as the enemy on the ground was slowing down our progress. I started ranting arrogantly, claiming that I could go alone and kill the enemy on my own, provoked by my frivolous, boyish anger. The sergeant turned to me and said in a mocking tone, "Well, why don't you go and get em' then?"

He turned his head to Unruh, "Unruh, take his place!"

"Roger, Sergeant!" I replied.

I got up and started walking towards the rear hatch. Unruh, took a widened glance at my advancement as we crossed paths after I jumped out the hatch. After taking the position on the ground, I turned around and asked,

"Can I take a battle, buddy?"

Shaking his head, my sergeant dismissed me and said, "Steele, go with him."

At that point, I was on the ground with Steele heading towards the enemy.

My friend, Unruh, who had taken my spot unfortunately, didn't make it.

Apparently there was a sniper at the top of the hospital. From where he was stationed he had a direct view of everything that was going on on our side. The bullet that ended up taking my friend's life had entered his body from the pelvic region, ricocheted, and had somehow ended up getting lodged in his heart. Although the medical team tried their best to resuscitate him, it just wasn't enough. Despite all their efforts, they couldn't figure out why the soldier was losing blood pressure, and it was soon afterward they pronounced him dead in the field hospital.

My friend, who died that day, hailed from Arizona and was a lot like me. Much like me, when he first came to Iraq, he found himself filled with fear. Slowly, as time passed, however, fear loosened its grip on him too, and it wasn't for a last-minute change; there was a high likelihood that I would've been the one wound up dead, while he would've been writing a book of his own.

To add salt to the wound, it was only after the mission was complete that I found out that my friend had died, and it had been too late. I couldn't do anything to save my friend. Still, I remember his bloody knee pad at the position where I was initially standing, which was all I saw of my friend. At that moment, it was almost as if time itself had ceased to exist. I couldn't focus on anything else.

I asked the soldiers around me weakly, "What happened."

"He was hit," I heard someone reply as the world around me got all hazy and disorienting. I knew that I was surrounded by uniformed soldiers and sergeants, my brothers-in-arms. Still, as I stared into one of my fellow soldiers' eyes, who, moments ago, had been smiling and calling me his "brother." I felt loneliness and despair in every inch of my body. Although I was safe at the base I felt despair unlike anything I had ever felt before.

Even now, as I think back on that day and the friend that I lost, I can't help but shed a tear or two in his fond remembrance. It's never easy to lose friends, but when those friends are lost in unfortunate circumstances, you feel like a part of you has died too and will be buried with them. While the smoke and fury of Iraq was a significant life event, a part of me still feels indebted to my best pal, my brother in arms, who I will never forget, with tears in my eyes and tenderness in my heart. "There is no greater love than to lay down one's life for one's friends" (John 15:13).

As far as Iraq goes, the incident with my friend was only one such event. Iraq changed me. This was my first and only deployment, but I found myself a different man when I left the country than when I first got stationed there. Still, as I write this and reflect on the life that I've lived so far I realize that Iraq, although a significant event in my life, was only a stopping point on my life's journey.

Throughout my life, throughout the ebb and flow of time, and all the places that fate or my choices have taken me to- all of it has to begin somewhere. For any life story to commence, there needs to be an origin point, a source from which all the fountains spring forth. For me, that source lies in my nana's house, where I spent a significant portion of my childhood and learned some of the most essential life lessons that I would carry with me everywhere, even as I fought as a soldier on foreign soil. Then, it is over here that I'd like to take a step back from Iraq and the monumental impact that it had over me, and rewind all the way back to the beginning.

Chapter 1

Beginnings and Childhood

Every story needs a beginning, a crescendo, and an end. While specific periods in our lives have a lot more impact on us than we'd like to admit, no one can deny the influence that their childhood had on the person they became as adults. Although my career as a soldier in the army would lead me to many places, the house where it all began was my nana's. Even as a grown man, I look back with fondness upon my nana's house and all the core memories that I made there. Although it wasn't particularly large or luxurious, the house always felt like home. Crammed with thirteen people, there was no shortage of love and tenderness in my nana's house.

I remember that, back in my childhood, there was nothing that I enjoyed more than sprinting in the cul-de-sac out front. I would sprint for no reason other than I wanted to do it, and I would even time myself to get as fast as possible. At my nana's house, sprinting was encouraged and it ended up saving my life in Iraq. I was allowed to keep up with my hobbies and was showered with encouragement for the most part. Thirteen people lived and shared their lives under one roof, and it was perhaps as a natural consequence of our living situation. We all fostered a loving and comfortable bond with one another except my uncle Mark. He was a bully. Among all the people living at my nana and tata's house, there were also my aunts, uncles, siblings, and of course, my mom and dad.

Although he played a major part in my childhood, my "dad" wasn't really my biological father. Despite this minor fact, he still treated me as though I was his biological son. As a result, I treated him no differently as I would my biological father. While we didn't share the same blood in our veins, we still loved each other and had each other's backs. In comparison, I've never really gotten along with my biological father. In the few rare conversations that I've had with him, I've only been reminded of how different I am from him. As far as I'm concerned, I've got only one father/dad; even though he isn't my biological father, he still loved me and took care of me as one.

My mom and dad had what you could call a whirlwind romance. They shared a few fleeting glances at a neighbor's party, and before either of them knew what was happening, they were on their first date. Something magical must have happened between them during their first date because soon afterward, they were in a relationship and living together. What started off as a whirlwind romance soon turned into a whirlwind life, since soon afterward, my mom ended up getting pregnant and giving birth to my siblings, Bubba, and then my youngest sister, Elise. Me and Gabriel, who was two years younger than me, welcomed two more siblings, and we couldn't be more excited.

My mother, who was originally from Inglewood, California, moved to Arizona when she was only two years old. She knew what it was like to be an outsider, and instead of giving up in the face of adversity, she took on the challenge. As an adult, she worked as a bank teller at Western Savings. It was a reasonable job, but my mom soon decided that she wanted more. One day, she woke up with ambition filling up her heart, and she decided that she was going to make a career out for herself- a career that she actually wanted. If this was any other person, I'd have my doubts about whether or not they'd be able to achieve such an impossible goal. But knowing my mother, I knew that one way or another she'd be able to get what she wanted.

Not only was my mother headstrong and dedicated, but she was also resourceful. Raising children alongside managing her career, my mom always knew which strings to pull to get the desired result that she wanted, and she'd make sure to do *just* that. No circumstance could ever bound her, and she made sure to cling to her beliefs and stand up for what she thought was right.

Truth be told, it wasn't all that surprising. My nana, who was one of the kindest women I knew, was also headstrong. She passed this trait on to her daughter, who, in turn, passed it down to me. I've learned from both my mom and my nana that if you truly have faith in yourself, you should do everything in your power and get to work to get to that goal.

It wasn't all just preaching. Faith without work is dead, from what I saw. They had faith in themselves and got to work on the goal. Mom and nana rarely ever preached. Instead of preaching, they defied their circumstances through their actions. Coming back to my mother and her monumental career switch, my mom decided she didn't want to be a bank teller anymore. Instead of staying in the same dead-end job for her entire life, she decided to enroll in an associate program for Accounting and got a degree in it. Once she had the degree, my mom got a better job at the Civic Crowne Plaza Hotel and thus began the formal inception of her career in accounting.

It was right when my mom got her new job that we moved out of our nana's house. The house that my family and I moved into would become the witness to some significant events that took place in our family. When we first moved, however, all six of us were gripped with a sense of accomplishment, and it was then that I realized for the first time that my life was going pretty well. A few things changed for my church visitations when we moved. When we were at my nana's, this old lady named Fay took me to this church with a tall tree, as tall as the church I loved climbing. But when we moved I met my mom's coworker's daughter. She was way older than me but started inviting me to hers anyway. It seemed to be God's way of keeping a presence in my life now that I look back at it. I was this little thug looking Mexican kid that slicked his hair back and wore Dickies, but she still offered to pick us up and take me and Gabriel. We went from living all in one big room to having our own rooms, we had cable as well as the internet. Now I had access to all kinds of music. The cable was a new thing to me, so I was awed to say the least.

Despite the sense of accomplishment and comfort that loomed over our heads, there has to come a point where cracks start to appear. Those cracks started to appear at Cactus Kids preschool before starting Junior High. My mom worked so much I was picked up by them from elementary school, and a couple of kids there began to bully me for no reason. The first couple of times that it happened, I decided that the smart thing would be to endure silently and let them have their way. After some time, however, I knew that I couldn't keep my anger inside any longer. One day, when the same group of kids came towards me with the intention to degrade and bully me, I decided that I had just had about enough of it. I was pissed off, and I knew that if I let them have their way, they wouldn't ever stop. So, that day, I fought back. To be honest, a part of me didn't think that I had it in me. After I had thrown the first, powerful punch back, I took a step back, with my eyes widened in shock and awe, and thought to myself,

Oh, shit!

After that, a smile broke out on my face, and I fought back as valiantly as I could. My fighting skills proved adequate since the bullies made sure to never cross paths with me again. My dad, perhaps, saw something in me, because soon afterward he enrolled me to box for the Maricopa County Sheriff's Department. There I learned to box and find an outlet for all the rage that I had pent-up in my hormonal teenage self. My first sparring match was pretty good. The coach pointed out the guy in the ring sparring and said I'm next. I watched him and noticed he telegraphed his punches and occasionally cocked his fist back too far before punching. I got in the ring and the buzzer went off. We started exchanging blows as the coach critiqued us. Keep your hands up and move your head, he says, and I catch that moment he cocks back too far. When he did I hit him with a hard left, he was stunned so I hit him again and again, so the coach stopped. He wasn't recovering.

I had a couple of fights under my belt, and I enjoyed boxing a lot. I boxed for the Maricopa County Sheriff's Department, and the colors we wore were black and yellow like the Steelers, which always made me feel some kind of way. One of the most important lessons that my dad taught me was to fight back. As a glass cutter, my dad was an integral fabric of the tapestry of my life, but as I grew up, my dad's drug problem escalated and got out of hand. Moreover, as his drug problem escalated and reached new heights, he developed a penchant for street life and often got into petty street crimes. Then, it's no wonder that he was in and out of prison a lot. As a result of my father's rapidly deteriorating behavior, he and my mom separated, and he was no longer a part of my life. I suppose I could remember him for all the problems that we'd later come to associate with him, but if there's anything that I've learned from my time as a soldier, it's to take in the positives.

To give credit where it's due, there were a lot of positives in my dad; he encouraged my boxing, and he was hilarious. I remember that he would often come up with his twists on words, where something like "Walmart" would become "Small Fart", and Sigourney Weaver he called "sick horny beaver."

He would put a twist on words all the time. He also switched letters and words: for example, toothpaste was "poothtaste" or "paste of the tooth," and my siblings and I would burst into laughter. He also had a penchant for watching the discovery channel with the utmost interest, and even today, whenever I watch discovery, I can't help but think of him. I never send him cards, but I text him happy father's day every year.

However, despite all the positivity, everything wasn't as smooth as you'd expect it to be. My school life was terrible, so I went to several alternative schools. First was Desiderata, Ombudsman and P.C.E for junior high, then West Phoenix for highschool.

I got kicked out of Junior High due to a certain incident. The incident, in itself, was quite hilarious. A class was going on, and I was prompted by the pang of hunger that had gripped me. It was after Halloween so I had plenty of candy. I decided to pull out candy from my backpack and eat it in class. Considering I had a knife, I knew I had to be careful, but it was a challenge that I was willing to take on.

With all the sleaziness of a teenager, I stuck a hand into my backpack and guided it towards a piece of candy. Although I had kept candy in the front pocket of my backpack, I couldn't seem to reach it, until I unzipped it and my hand just slipped into my unzipped bag with ease. My teacher was walking by and was opening unzipped bags to check out the candy out of curiosity then came up to my bag and said "so what kind of candy do you have?" As she flipped open my bag and saw an object that she said was "not allowed on campus."

Much to my dismay, the object turned out to be a pocket knife, and judging by the sheer look of horror on the teacher's face, she wasn't particularly keen on me bringing a knife to school. What sucked, however, was dragging me to the principal's office.

As a consequence of the knife/candy incident, I was kicked out. I spent most of my time in alternative schools.There's one thing that you need to know about alternative schools, and it's that you're exposed to a whole arsenal of strange and weird people at a very young age. If there's one thing that the entire experience taught me, it's that people are just downright bizarre. Still, despite all their strangeness, you can't help but be drawn in.

Chapter 2

Childhood

No matter where you're from, or the places that you're bound to end up at, your childhood defines- in more ways than one- the person that you're going to be. The experiences that I've had in Iraq were to leave a solid mark on me for all my life to come. In order to understand how I ended up in Iraq, it's worthwhile to turn back time for a bit and start at the very beginning. While I've a glimpse into the life that I had as a child, I'm going to take a deep dive into my memories and produce as accurate an account as I can of the days when I was growing up in my nana's house.

Memory works in mysterious ways. Today, when I sit down to write about the days of the past, I feel as though I'm trying to collect running water in the palm of my hands. To put it rather simply, there's just *so* much to talk about. No matter how much I try to keep the water in my hands, I know that despite the best of my efforts, there will undeniably be an instance or event that I'll forget to talk about. While that may be true, I don't think it takes away from the experiences that I do remember in any way whatsoever. If anything, in the palace of memory, events- while they may appear like that at first- are never stand-alone. One instance links inexplicably to the next, and long before you know it, you can recall with vivid clarity the days of the past. That is precisely what I intend to do over here. Although I know I can't actually turn back time and go back to the days of my childhood, in writing about them, I'm hoping to pay tribute to the experiences that shaped me into the man that I am today.

Anyway, to start at the beginning, I need to return to my nana's house. I started junior high when we moved out of my nana's house, but despite that, it still took me some time to blend in. Much like every other kid who's moving to a new school and acquainting themselves with new surroundings, I felt a strange mixture of emotions. A part of me was overwhelmed, while another part of me was excited and looking forward to what was to come. To further add to the overall newness of those days, my mom was moving up into an accounting career and had just gotten her hands on a much better job. In those days, my hobbies depended entirely on the availability of my friends. My mom had me and Gabriel in karate, and besides that, I did break dance whenever I got the chance. For the most part, I spent a great deal of my time with my friends, and perhaps the explanation for that was the fact that all my friends were kids in my neighborhood and junior high. Most of my friends were practically neighbors and the rare bunch that didn't lived within the radius of my high school. Estrella, our junior high school, was next to El Oso Park, which was also next to Trevor G. Browne High School.

I remember a lot of the interactions that I had with my friends since I met these kids at a time when I was just beginning to come into my own. I was easily impressionable, and boy did my friends leave their mark on me. Looking back on the whole situation in retrospect, I don't mean that in a metaphorical way. Richard, a friend who lived near my school, actually managed to leave a literal, nasty mark on me. This was around the time when I was preparing to go to junior high, where I met all my new friends. Right before we moved into our own house, and were still living at my nana's, my mom would often get me model rockets.

At Junior High, most of my friends were of the rare group, and I suppose that I picked up on some idiosyncrasies. Cheryl, Robin, Danielle, Matt and Melissa were among my first proper friends in junior high, and I loved spending time with them. Perhaps, it was because of my interest in model rockets that I wound up with Richard launching mortars in the air... Anyway, let's get back to Richard and rockets. To be quite honest, Richard was kind of a bad kid. Well, he wasn't "bad" in the general sense of the word. I mean, we didn't go out and get drunk or anything like that. Instead of drowning ourselves in alcohol, we would go to a park, mess around with the rockets, and launch a couple of them.

One time, we went to the El Oso park to launch a bunch of F5 rocket motors, which included some of the biggest ones that I had. After we'd managed to sneak up to the park, I would hold the pipe, and Richard would hit the button. The rocket would be all hooked up, and then we'd launch it off into the distance. Most times that we did this, I thought it was really fun. Usually, we'd burst out cheering every time the rocket launched, but just this once, things went terribly, terribly wrong.

I was lying down, holding the pipe, and I don't remember what it was, but for some reason I burst out laughing like a little kid. Richard hit the button, and I kept staring at his face. In that moment, I was on the floor, smiling like an idiot. Finally, I looked at the pipe, and then the realization hit me- the motor was pointed right at me. In the next instant, before I could even open my mouth to let out a single word, the rocket went off and shot right at me. In my naivety, I thought that it wasn't that big of a deal.I popped up on one arm thinking it passed. I popped down right on it and when I got up it was in my shirt. Trying to get it out and burning my rib, the rocket finally found its way out as I flailed about, yelling "holy shit and took my shirt off!" I turned to face Richard and saw that he was turning white.

"It's no big deal, dude," I said to him casually.

I looked down at my rib and saw that it was all charred and black. The pain stung me wildly, and I let out a muffled scream.

"Oh, no!" I yelled.

Exasperated, I didn't know what to do. I knew that if my mom found us out, she'd kill me, and in my desperation, I couldn't see a way out.

"Don't worry. I've got a neighbor down the road." Richard said as he tried to keep his cool. "He'll probably have the medical supplies we need to fix you up."

With that, Richard flung his arm underneath me, and we made our way to his neighbor's house. We started by taking small steps, but with every damned step that I took, the pain just accelerated. Infuriated by the intensity of the pain, I made a run for it. It was a busy street that I had run through, but at the sight of an injured kid yelling and running, everybody just hit the brakes and stopped. I crossed the street and ran into the neighborhood. Finally, after a dangerous journey, Richard and I went to that guy's house. The man didn't say anything, but he gave us antibiotics, bandages, gauze, and whatever else he thought might be of help to patch me back up.

The man then had me lay down on his couch and patched me up. Still, he said that I would need to go to the hospital to make sure that the burn wouldn't get infected. It was at that moment that I remembered that my aunt worked at the hospital.

"My aunt will have supplies, right?" I said to Richard, and he agreed.

We went to my aunt's room.

"Do you have antibiotics and some burn ointment on you?" I asked her, trying to keep my tone as casual as possible.

Immediately, my aunt could see that something was up, and inching closer to me, she asked,

"Why?"

I showed her my burn, and her face contorted into an expression of horror.

"Oh my gosh!" She yelled. "Belinda! Get over here!" My aunt called out to my mom, who was in the other room, since at that time, we were all still living at my nana's still, and she immediately rushed to the hospital.

They ended up cleaning it up for me and had to use a wire brush. I don't remember much about that dreaded hospital visit, except that it hurt a lot more than I had anticipated. In the end, however, they got all the dead skin off. Before discharging me, they gave me antibiotics and ointments, and as a result, the burn healed pretty well. Today, if you were to look at the spot on my body where Richard shot a rocket into, you'd be surprised to find nothing at all. Although it was a third-degree burn, it's almost as if nothing ever happened, since there isn't even a faint scar.

Soon after that incident, however, Richard and I were caught, and that brought about an end to our crazy rocket escapades. While I enjoyed hanging out with Richard a lot, I did have other friends that I hung out with at my school. Most of these friends were girls, and I had to come right out and say it: I really liked hanging around girls. I mean, there were occasions where we fooled around a bit, but for the most part, they were pretty cool. Cheryl, Robin, Danielle, and Melissa were the first proper friends that I made at junior high school, and Cheryl and Robin were twins. We'd actually go hang out at their houses and just chill. There was also this kid, Matt, who'd come around and join us. There was also another guy, who I remember being called "Tall Man," and we would often go over to his house and chill. Tall Man was luckier than the rest of us; for some reason, he never had parents at his place. As a result, we almost always went over to his house, played video games,brought the girls over and jumped on the trampoline. Basically, we could do whatever we wanted, and as young high school kids, that sweet taste of freedom was unlike anything that we'd ever tasted before.

I'd hang around the girls and Matt as much as I could, but after some time, we moved. We no longer lived at my nana's house, and my mom purchased a house of her own, which just so happens to be the house that we live into this day. Now, I still hung out with the girls, my friend Matt and a couple of other people. But as a result of the shift, my group of friends kind of changed when I got into the "incident" with the school. As a result of my little misdemeanors, I got into alternative schools, and my set of friends completely changed. I wasn't hanging out with Richard anymore, simply because he moved. Still, I tried to keep in touch as much as I could, even when I was going through alternative schooling.

My old friends, however, were going through their own changes. They started hanging out at Melissa's sister's boyfriend's house, who was 21, and one day, they invited me too. There were quite a few of us, including Pablo, who was 21, Melissa, Robin, Cheryl, and Danielle. We just hung out for a little bit, and then everyone else took their leave, except for Pablo. I went from hanging around with my friends to just hanging around with Pablo who was a few blocks away.

The first time that I got drunk was with him. He was 21, whereas I was only 14. I remember his mom getting pissed that he was hanging out with kids, and soon afterward, that came to an end. It was at that point that a series of breaks and interruptions took place within our friendships. At the alternative schooling, I had to face a lot of difficulties that the others didn't really understand, and soon enough, I started hanging out with the girls and Matt less and less. There came a point when I wasn't hanging out with anyone at all, and it was at that point that Gabriel invited me to hang out with him and his friends. This is the point when everything starts to get really *interesting*.

When I started hanging out with Gabriel's friends, Gabriel's group was two years younger, except for Clint's brother Stuart. It was almost as if I'd unlocked the door to a secret world somehow. There were a couple of reasons for this, but the most obvious one that I can think of is age. Everyone in Gabriel's friend group was approximately two years younger than me, Willis, and Clint, except for one person who was my age was Stuart. Suddenly, in the span of a couple of weeks, as I started hanging out more and more with Gabriel's group. I started to pick up on their habits and became like one of them overtime. All these kids, who were younger and ten times cooler than others I hung with, introduced me to weed, drugs, and some great music. One of the memories that I look back the most fondly on was being introduced to Tupac. I remember the first time I listened to Tupac's "Who Do You Believe In," and I felt goosebumps crawl all over my skin. To this day,

I'm a huge Tupac fan, and if it wasn't for Gabriel and his friends, I might have missed out on some of the greatest music that I've ever heard. I would mostly listen to DMX, Snoop and DJ Quik because I heard my dad listening to it and it sounded great.

But, of course, the experience of hanging out with a different group of friends has its disadvantages, as well as its perks. If the perks included great music and weed, the downside of that whole ordeal was learning how to deal with the "thuggish" nature of my newfound friends. Strangely enough, the best way that I can think of to describe it is as a dark and heavy cloud that loomed over some of the kids, and as their friend and confidant, you had no idea when the cloud would pour, and the thuggery would take over. Truth be told, I wasn't really into the thug lifestyle. I mean, I knew how to hold my ground and could throw a decent punch, but apart from that, I liked to be left alone.

As a young teenager, I was a firm believer in the "live and let live" doctrine, and I didn't go chasing after the thrill of a fight. Some of my new friends, however, had a penchant for this kind of stuff. Not only would they go around wreaking as much havoc as they possibly could, but they also went around town committing a series of petty, and now that I think about, *childish* crimes, such as stealing shit from the grocery store, amongst other things. Once they'd gotten bored with stealing, they'd pick fights, and most times, this would result in two possible scenarios. In the first scenario, the other side wouldn't budge and would simply ask us to piss off. Alternatively, the situation could also escalate, and the whole group would have to orchestrate an entire group fight, which in my opinion, was too much of a hassle for an altercation between two kids. But in those days, you just had to go along with the group and stand up for them.

In those days, being a couple of teenagers, we were always chasing one thrill after the other. There came a point where fights were just not cutting it anymore. It was during this point that I tried acid for the first time. During that time, we would hang out at Willis's house quite often. Looking back at it, it was one of the high points of my teenage years. The whole group would assemble at his house, and we'd chill out, smoke some weed, invite some girls and have a good time overall. You know, we just enjoyed hanging out with each other, and once Gabriel had introduced me to the group, everyone treated me as though I was their best friend. I tried the little magic square and went on my first acid trip. The experience was, as you'd expect it to be, trippy and absolutely batshit crazy.

Considering how the acid trip was one of the most profound experiences of my youth, let me paint you a more accurate picture of how the ordeal had gone down. The group had assembled in our usual hang out spot, and after we'd chilled for a couple of hours, Gabriel turned towards me and asked casually,

"Hey, do you wanna do some acid tonight?"

I thought about his offer for a moment. I had never tried acid before, and I had heard faint stories about people having a "bad trip," but I was too curious. I wanted to experience a trip for myself, and I shrugged my shoulders nonchalantly and answered,

"Yeah."

"Do you have nine bucks?"

I shuffled through my pockets and found a crumpled ten-dollar bill.

It was during this time that my mom had us move into the new house. We were getting allowance for chores weekly, and as compensation, my mom would pay us money. As a result of my helping around the house, I had money. The new house afforded me a lot of new incidents and events.

I handed the bill to Gabriel, and both of us got acid. Later that night, we pulled the little squares out and placed them on our tongues. What happened next occurred in the course of a few hours, during which I was sitting in my friend Willis's room, staring at his dog. After two hours of taking the drug, I kept staring at his dog and found the pit bull staring back at me, looking rather bored. At that point, I had waited for almost two hours, and I had just about had it with the waiting.

"That was a waste of money," I mumbled to myself, and it was as though the drug was waiting for me to say the words to activate itself.

The dog cocked her head to the side, and all of a sudden the background around the dog turned a vivid shade of purple. After turning into a bright purple, another shade of color approached the creature, and then the color turned green. The whole time, the dog stayed the same color while being approached by a kaleidoscope of different colors, varying shades of red, yellow, orange, and blue. Staring at the dog, I grinned wide.

"Oh my god! This is actually working!" I thought to myself and smiled some more. "I'm tripping," I said out loud, but there was something wrong with my voice. The words came out of my mouth at a much slower speed and rang in my ears as though they were being emanated through a warped loudspeaker.

The weird kaleidoscopic trip that I was on went on for a couple of hours, and after being bombarded with a plethora of colors and shades, I felt my eyes growing weak and my thought process going fuzzy as the clouds seemed to move so fast.

"This might be a little too much now," I thought to myself, and the thought rang violently in my mind. Exhausted at the vibrancy of all the colors and shades, I kept my eyes glued to the pit bull, and when I couldn't do that anymore, I thought out loud.

"I should close my eyes."

I blinked a couple of times but didn't seem to be able to close my eyes.

"Let me close my eyes."

After trying to close my eyes, I finally succeeded, but much to my surprise, it was even worse with my eyes closed. I could have tried to open my eyes again, but it had taken so much work to get them to shut in the first place that I thought that I'd just go along and roll with the blasting colors as they came. So I began to look at the clouds which were moving so fast in my eyes and I calmed down even more.

After some time, I heard a slowed, weird voice speak to me from a distance that I could not comprehend. Maybe they were near me. Maybe they were far away. At that moment, I couldn't figure out where the voice was coming from, and I didn't really care about it.

"Daniel, do you wanna go mess around?"

I must have answered "yes," because the next moment, I was across the street at the store giving the bum money to get us beer. I must have blinked because when I opened them, I was on another street, where they were constructing new houses. I saw someone who looked like Gabriel and my friends break a couple of porta-potties, and I thought I might as well join them. In the next couple of moments, I remember that I fucked up a couple of porta-potties and kicked some beams, just for the hell of it. Time moved slow and fast at the same time, and it was hard to keep up with what was going on, but caught up in the movement, I caused as much damage as I could. My friends did the same, and I guess someone wrecked the big main roads that were down the way, and I blinked my eyes again, and suddenly, the red and blue of police lights were reflecting on the wood, the porta-potties, and the beams. Tripping as I was, I could only see the blue and the red, and at that moment, it hit me.

"Oh my god, we're surrounded."

I grabbed Gabriel's hand, or I tried to, and I yanked him towards me.

"Police, come to get us." I tried to explain, but he just kept giggling.

"No laughing." I pleaded with him and kept yanking his arm. "Come on, come this way."

Everyone started following me, and even when I was blown off my knockers by the acid, I still managed to make everyone follow me. I was the first one to jump off the wall, onto the other side, into one of the concrete canals that were used to pass water onto farmlands. My feet hit, and they flew out from under me, and I just fell straight into it, and then my butt hit. I folded up like an accordion, and then I popped right back up. My back was messed up, but I didn't have time to feel anything. In the acid-induced panic of the moment, I just popped right back on my feet and kept running. I ran into something, and then *boom!*

My right shoulder hit something as I heard a neigh, and I felt something with my left shoulder. I heard hooves, and I thought to myself, "What the hell?"

It turned out that I had just hit the fence surrounding a stable. This must have been electrified or something because when I popped back up off the ground, my arms locked up. I guess it was because I was getting a little shock or something from the fence. Once I had registered the shock, I got up and started running again. When we got to the street to cross it, we saw that there was a giant wreck. When we finally got back to our senses, we found out that there was nothing to worry about, and we'd gone through all of that for no reason at all. All of us had had a good laugh at that, and I vowed to never try acid again.

Apart from the acid incident, there was also the time that we used to steal a truck. Oftentimes, when we'd hang around Willis's house, we'd drive into the desert, blast out music on the stereo, and listen to some Tupac on the ride home. Once, we stole the truck, and we were out driving it in the desert. I'm pounding on the hood, standing up, doing all sorts of batshit crazy shit. In those days, we'd go out there to the desert, and you know, just fill bonfires or just do crazy stuff in the car or in the truck. Back to my incident, however, as I said, I was standing up, banging on the top of the truck, and I was yelling and screaming, half out of my mind. At that moment, the driver hit a bump, or he took a little turn or something.

One moment, I'm acting batshit crazy, and the next, I go flying out of the freaking truck, head first. My head hits, and my left shoulder hits, and then I just start this crazy tumbling. When I stop rolling and tumbling, I'm sitting straight up, and I'm looking at the back of the truck. I don't know how I'd landed straight up, and I was too dumbfounded to form coherent sentences.

"Whoa," I mumbled.

My friends got me out and looked at me with their eyes popping out of the socket.

"Dude, are you okay?"

"Yeah, I'm fine," I replied, looking dazed as all hell.

"You sure?"

"Yep," I said, clearly shaken.

"Hop back in the truck then, homie."

I got up and sat in the stolen truck. This time, however, I made sure to act like a human being, and as the truck burst into movement, I slowly fell asleep. Looking back at it, I can see that that might not have been the best move to take, but at the moment, I was just relieved for a few moments of rest and relief. Once I'd reached home, and the bruises on my back and shoulder had healed, all my friends had a good laugh, and ever since then, I made sure to never go batshit crazy in a moving vehicle again.

Chapter 3

School Days

It was time to pack our stuff and shift to our new home. It was not only exciting but also a saddening moment both at the same time. We didn't want to leave Nana's place, as it was so much fun there. My mom would take my siblings and me to different parties with her. Those parties and the people there always fascinated us. We would run around everywhere, while middle-aged men and women would be dancing to some catchy beats like Dr. Dre, The Chronic album, and other albums like it- even mariachi music at some. Some parties had no music, just a fire, so those ones would be a little boring for us.

However, my mom enjoyed every moment of her life to the fullest during that time. She was all carefree and content. Even though my father was in prison, unfortunately, my mother never made it a reason for her and not to have fun. She spent the days at my nana's like and she was a cheerful and resilient teenager at all times.

However, once we shifted to our house, she was no longer the same. Maybe she had more responsibilities to fulfill, or she missed my father. Soon enough, however, my dad was out of prison and he was a part of our lives again. He showed up and said "we are Steelers fans now," so I looked into them and decided to be a fan at the age of 15. It was just then she got busy with her job as an accountant for the Civic Crowne Plaza Hotel and with the household chores. Also, my school was about to start, so maybe that was on her mind too.

I was supposed to start high school at Trevor G. Browne, but because of my fighting, I was suspended. They wanted me to repeat a grade because I missed out on so much the first year by getting in trouble, but I ended up going to West Phoenix High School instead. This was where Gabriel, Joe and I were in the same grade. There were mixed emotions. I was both excited and nervous at the same time. There were a lot of thoughts crossing my mind, like will this school be like PCE, good or bad? I thought a lot about this transition from Starlight Park Elementary School to Estrella Junior High School, but it had to happen.

I believe that school days are the best time of our lives. We are doing our own thing without any problems in the world. We get to learn so many new things and also make amazing friends. Even I met some great people during my time in junior high, though this was only a year, but I have thousands of memories to cherish forever.

After school I would hang out with Robin, Cheryl, Danielle, and Melissa. The twins and Danielle have been my friends since junior high. We would attend social events at school and would also go out and hang together. However, this all ended after six months when Richard moved out. I would like to clarify, however, that these were two separate events that just so happened to occur at the same time. Richard had nothing to do with the girls not hanging out with me any more; I don't think he even knew them for that matter. Going to these different schools pretty much did it.

Even after that I was able to find people who made me experience some of the wildest things in life. This takes me to the day when I tried weed for the very first time. There was a dance in junior high, and from what I recall, it was as most dances were back then- highly chaperoned and *boring*. I was just walking around on the sides. That is the moment when I heard a voice from behind, and I turned.

"Hey, look, let's go sneak out. And we'll be back."

I didn't know this guy; I had never seen him. I turned around, and saw that my friends were with him, and all of them seemed to be in on the plan. Well, I sure as hell wasn't going to be a party pooper.

"Okay, yeah, let's go," I replied and followed the group.

So, we snuck out and went to behind Circle K. This guy pulled out a bag of weed and started breaking it up. The whole ordeal was quite new for me, and I was completely gripped. I kept my eyes glued to his hands as he broke the weed into smaller, smokable pieces.

"What are we going to smoke out of?" I asked.

He took out a beer can and crushed it up with his hands, taking great care to ensure that he didn't crush it entirely. Once the can was sufficiently dented, he poked about six holes in the can, and placed the weed inside it. The guy smoked through the opening of the can, and passed it to me, urging me to do the same.

It was my first time smoking weed. I had read at multiple places that weed does not get you high when you smoke it for the very first time, but I suppose in my case, it was an exception. I got extremely high after smoking only a single joint of weed. I never knew that it makes you feel over the moon. It felt as if I was the happiest person on this earth, and I wanted to make the most of it.

My heart was pounding way too fast. I could even hear the beating in my chest. There was no chance that I could dance in such a condition, but Amber grabbed me and took me to the dance floor.

"Oh yes," I shouted.

"I'm so fucking high. I can feel my heart beating in my chest," I said, and she started laughing.

The night was perfect. I smoked weed for the first time and had my first dance. It felt as if I had conquered the world. We just kept dancing and the night went on.

Many of my friends were going for an after-party they called a kick back, but I was too tired to join. I just walked home and landed on the bed and passed out.

* * *

The science class was my favorite. I would love to attend my science class only, because of Miss Moon, who was an awesome teacher. She would teach us autopsies and dissections that were quite interesting, and these dissections were carried out on frogs and even squid. One day, Miss Moon gave one squid to each of us and asked us to cut off the arms and chest of the squid. She also brought a hot plate to cook the tentacles and told us that technically it was calamari, so I tried it of course. It tasted like chicken. Although it was quite disgusting, it was a great learning experience.

One of the reasons why Miss Moon had always been my favorite teacher was that she made science fun. The subject that makes children run away was my most liked one, only because I was experiencing science practically. She would not make us write papers on how the universe functions, but would make us watch videos and documentaries to improve our understanding of it. Apart from science, the cooking class was also one of my favorites. In junior high, there was a separate class dedicated to cooking where we got to cook different kinds of brownies, cookies, pies, and pastries called home ec.

Apart from some happy moments, there were some frustrating moments at junior high as well. It was when I tried football in school, and the football team had a lot of rules. I was interrupted for not having the required accessories with me, and every time they told me about a different accessory that I had never heard of. One day we had a match, and all the players were going head-to-head and tackling each other, but they didn't let me do it. Upon asking, I was told that I didn't have a mouthpiece which is why I was not given the chance to tackle head-to-head. All this made me very frustrated, and I left the team then and there. However, I was not in the school until the eighth grade, so my football experience was done.

I stayed at Estrella Junior High for only 3 months, and after that the knife incident happened. I was carrying a knife in my school bag, and I got caught by my teacher, which led to me being expelled from the school. However, my reason for carrying the knife was very simple. I carried it with me for protection, and most times, it was just easier to show any bullies a flash of the knife, rather than taking the trouble to get up and smack them upside the head. I didn't want to hurt anyone or injure anyone from that knife, and it was just that I wanted to make things a bit simpler. I guess I looked somewhat like a cholo with my hair slicked back like some greaser from *The Outsiders*. However, this outlook on life and that knife cost me a lot.

Then I got enrolled in another school named Daisy Desiderata. This school was a school of bad kids who would not pay attention to the teacher and would mess around in the classroom. They were nothing like the children of my previous school, which is why it took me quite some time to gel in there. One day my class teacher came to me and said that I was different from all these kids as I was a lot more well-mannered and disciplined.

When they all would be bouncing around and acting crazy, I would just be in my world doing my work or drawing. They didn't have my attention at all. Another troubling thing about this school was that it was on 16th Street, a bit far from my home. I had to walk past a pretty bad neighborhood in order to get to my school. No matter where I got off the bus, I would still have to walk through that part of the neighborhood where all the crackheads lived. They were basically drug addicts who consumed fentanyl, crack, and methamphetamines.

It was because of these people that I had to leave my school again.

The knife never left my side, and it always made sure that I got kicked out of my school for doing absolutely nothing. So, what happened was that these areas were very dangerous, so I decided to keep a knife with me for protection. But little did I know that a knife would cause me trouble once again. But this time, the knife was small, just like a switchblade, but still, I got caught.

I was leaving the classroom when I heard my teacher calling out my name. I turned back, and he said,

"You know, Daniel, you dropped something over there."

I replied, "No."

Then he picked up the knife and said, "I am sorry, I have to send you to the office."

"Okay," I nodded.

Minutes later, I was in the principal's office where my termination was being drafted. As you can imagine, the knife dropped out as I got up from my desk. So I got kicked out of that school which actually relieved me because it was a crazy, *Dangerous Minds* type of school, if you ever saw that movie. I started going to Ombudsman to finish off seventh grade, which was very laid back with no real dress code or strict guidelines, but we did our work and learned very much. We learned about government history, and how laws are made and passed. It was like a requirement almost no other school had. t This school is where I bought weed for the first time. I knew the guy that sold it usually would bring it to school with him, so I simply told him I needed a nickel bag. After class he met up with me and five bucks got me quite a bit. I remember thinking damn, I can't smoke this all by myself, as I walked home from 67th to 79th Ave. It was Friday, so I decided to stop by this girl Jessica's to let her know I planned on coming by later to smoke with her. I went home, took a shower, got something to eat, and headed to Jessica's.

Once I got there we chilled in her room while I broke it up. I was still a novice pothead so I used an empty Coke can turned into a pipe. We ended up smoking all of it and I was too high to walk home. I called my mom to tell her I was spending the night at a friend's, thus concluding my first time buying weed but not the last. I easily passed seventh while still boxing for Maricopa Sheriff's Department despite my dad going back to prison. My coach Walt Hoskins lived in my neighborhood, so he would pick me up every Monday, Wednesday and Friday.

In the summer I would walk to his house to get a ride until he had a heart attack that summer, but he took me to Ombudsman. I was tested for learning disabilities for some reason, and it turns out I have or had ADHD (Attention Deficit Hyper Deficiency), so it was decided that for eighth grade I would go to Phoenix Center of Education. This school was far away, but my mom dropped me off and I didnt have to find my way back anymore. They took everyone back home. I didn't like the fact they had staff there specifically to restrain, you which I thought was stupid until I saw the students. They were disabled with rage problems and other things like OCD, but most of them acted straight crazy. The ones that chose not to be shuttled home often told crazy stories about them hooking up and doing sexual things in alleys and stuff. I got shuttled, often being offered oral sex by quite a few girls. I didn't last long at that school.

After about two months I was expelled. I was debating with the teacher about something and she was losing so she sent me to the time-out room. This is a room that you go into yourself or the staff puts you in there and they lock it. After 45 minutes I got very aggravated and started asking to be let out. Eventually I started kicking the door, so the staff came in to restrain me, taking me down and sliding me all the way across the room. The baseboards of the room stuck out a half inch and were unusually high, so my left shoulder blade caught on it, dislocating it momentarily which really hurt. I told him to let me go cause I was hurt. He didn't listen so I hit him three times with my right, immediately swelling his face. He didn't press charges, but from then on I did my work from home and they sent it to me.

At this point I started hanging out at this guy Ryan's house briefly on my Nana's street. He had racist skinhead friends that would talk shit here and there, but he knew I could hold my own, and it was obvious when I talked shit back with ease. I snorted some coke and meth and also took mushrooms just to try it, never making it a habit like they did. The music was different like Carry on Wayward Son by Kansas, Blue Oyster Cults' (Don't Fear) The Reaper, Foo Fighters and many other groups and songs-even Korn. We would often talk about the crazy stuff I did, like building a quarter pipe against my nana's brick wall and accidentally launching my bike over the wall, and many other crazy stunts I pulled. I was introduced to the game Metal Gear Solid tactical espionage and Mortal Kombat 4, and it was great. It didn't last too long, he ended up moving also but before he did I played the hell out of his drum set.

Winter came around and I developed an upper respiratory infection that turned into bronchitis, then the flu, and ultimately pneumonia. It got really cold and when I went outside my lungs froze up or something and I could barely breathe. My mom rushed me to the hospital where I stayed and recovered on some type of ventilator that cleared my lungs up till it was time to go home. Eighth grade was interesting now that I look back on it. Many things happen all in nine months or so. The summer before high school I had my first girlfriend. Sure, I fooled around with plenty girls but no relationship or sex until her. I was technically a virgin before then. I was more of a summer fling for Marlena and I realized that when she three-way called me, introduced me to her friend and broke up with me, leaving me on the line with her friend. After a while I realized I was handed off to this girl, and I told her how messed up that is and I'm not interested.

As for high school we will start with West Phoenix since I covered what led me there already. My mom would take me, Joe, and Gabriel to school and we would catch the public bus back and get off in front of Toys R Us. We would walk into "the hood" as we called it. Going to this school was not bad. Me, Gabriel, and Joe were in the same grade and had a few classes together. I met a few girls that I invited to Joe's here and there, eventually finding a way to my house down the road a few blocks. We bought lunch from the food truck. We had no problems. I stopped carrying a knife cause I was never on my own; we were a small gang and we claimed west side Phoeniquera. We were asked to write a quote we knew so wrote "to live is to suffer, to survive is to find meaning in the suffering" (Nietzsche). At the time I just heard it in a DMX song.

All was well until one day Gabriel bumped into this black guy who was apparently having a bad day so he started talking shit to my brother. Every insult was racist, calling him wet back, beaner, Jose, Pedro,

Mexican jumping bean and even dirty Mexican, as if he was allowed to. My brother just fired right back without being racist, but after a while I guess he ran out of things to say so he called him the N word-not nigga which is not a cool word either. We used the word ninja, which I started and it caught on amongst us. It was time to go to class, and Joe and I stepped in and told him to leave. Next thing we know, we are done with school and on the public bus standing up in the front, and Joe taps me and says look towards the back and there is that black guy and half the public bus standing with him. Apparently he went around telling people stuff about us, getting them all pissed at us, but I didn't know until later.

We got off the bus, and the guy and all the kids got off with him. I thought maybe these 15 or 20 kids were watching. This guy got in Gabriel's face, trying to make him flinch, saying things like you want to talk about my peoples Jose, continuing his rant. Then some guy punched Gabriel from the right and collapsed. Sincemy brother's face didn't budge at all, I just laughed at that weak punch, and Gabriel ignored the punch and started fighting the one he was facing in the first place. As the fight went on, they ended up rolling around on the floor, bumping into a guy and making him drop his CD player. Then he started kicking my brother. I ran over there and with one left hook, I laid him out and all hell broke loose. These people were not there to watch. We had to fight four to five or more each, and it took one or two punches from me and they were done. I finished my little group and turned just in time to see someone trying to sneak up, so he crossed his arms, put his head down and tried to plow through me. I simply put my left hand out to stop him and gave two upper cuts with the right. I had nobody to fight so I looked at Joe as he took his knife out, because they tried shoving him in a bush face-first then ran.

We cleared out everyone, and it was just Gabriel, but the group beat him badly. As the guy ran up on him, my brother kicked him in the balls from the floor and it was over. The cops showed up as I was helping Gabe home and he tried interrogating us on what happened. I told him and showed him Gabriel's face. He offered me a ride but I don't like sitting in cop cars. Other than that event, we kept going to West Phoenix and nothing else happened. I guess they knew we were not some people to mess with or they would be seriously hurt. I had to leave and attend Trevor G. Browne for the end of my last year of high school and took a computer class to have enough credits to graduate, but I did it easily. This separated me from my usual group but I went by fairly quickly, and before I knew it was time to graduate, cap and gown and all. I wouldn't have had a ceremony at an alternative school, that's for sure.

Chapter 4

Hanging at Joe's

When Joe, Gabriel, and I started going to West Phoenix High School, we all started hanging out at Joe's all the time. Before that we hung at Willis' or Clint's after school, on the weekend, and often passed out on the couch instead of walking home. Everyone would chill there: me, Gabriel, Willis, Clint, and many more, as well as these two girls Tricia and Candace. Though many came, these two hung out with us and stuck around as we partied at his house. Much changed when we moved. We went from living all in one big room at my nana's to having our own rooms, and she no longer used food stamps and, but she still went to the food bank and still does to this day. On the weekend I would take some extra food from the food bank to his house to grill up, and he would throw some stuff in too as we drank our forties and partied.

We might as well have been listening to that song Homeboyz by 2Pac every night. We funded our lifestyle with occasional beer runs and late night roaming the neighborhood, looking for things to steal. Stereo decks, speaker systems and even a few rims is how it started since people always seemed to need them. After a while we got better, and it seemed so easy to guess what cars had worthy items. Christmas time was especially profitable because presents would be in the garage or cars. It's messed up, I know, but we were prepared for the consequences. We called it "putting in work," but looking back now I call it stupid shit. We are lucky we got away with it. This one time we found more than a few people could carry, so we got the truck we often stole to move it to Joe's. It became second nature or routine I guess you could say. We often got together and we chilled inside, out front, or out back around the fire pit we dug up. Nothing but weed and beer for me, though they did acid a few times, but I was done with that stuff. The girls Tricia and Candace were tripping and jumped into Joe's nasty dirty pool, and I had to hose them down, eww.

On nights that we planned on roaming the hood there was no loud music and no partying, just a bit of drinking. Our night escapades must have gotten some attention, because the cops came on a night we planned on going out, and of course we smelled like alcohol. They brought us in for questioning and of course they got nothing from either of us. We all saw *Goodfellas*. We knew the two rules "never rat on your friends and always keep your mouth shut," so we told them nothing and they got nothing. Boy, when we partied though we had plenty of stereo equipment and it got loud.

Plenty of events happened when we started hanging at Joe's. The first was that huge fight in front of Toys R Us, which is now a grocery store on 75th Ave. Second, I was walking into my neighborhood and saw these three adult guys from sur threse, meaning 13th Street. A California gang or something started beating up Stuart, so I ran up, not to fight cause they most likely had guns, but I had to do something. I said "you don't have to do this," and immediately one of them turned to me, grabbed the back of my head, and shoved his fist into my face, breaking my front tooth at the root, but it was still there. I wanted to fight but heard the guns cock, so I just yelled at him, pointing at my tooth, saying look what you did you fucking asshole fuck, fuck, fuuuuck. They looked confused and I just turned around and went home, and my mom took me to the dentist. Now I have a bridge covering that gap, and it is a permanent reminder of that day. I have to be sure to floss under it everyday, so it reminds me everyday that people are the most dangerous thing on this planet.

I didn't go out roaming the hood with them every night. Sometimes I would stay and keep the girls company, and I lucked out when I didn't it seemed. I would look up and see the helicopter shining its spotlight, and I knew what was going on but they never got caught, and neither did I. "Started off runnin' from coppers, duck and hide from helicopters we open fire, who can stop us"(Tupac's "Lost Souls," 1997). That's just how it was if you hear the song. At one point, Joe let this drug dealer Johnny, from some gang called Califas, rent a room. So, when we needed weed, he was there and ready to supply the goods. Although we had many problems with the Cali gangs, Johnny and his brothers Bobby and Rudy were cool as hell. We referred to the Cali gangs as Cali bastards, but not these guys. One night Joe's brother went to the store, and when he got back he told us a group of guys chased him. Me, Joe, Rudy, Willis and Clint told him to take us there, and when we got there the group was three guys. Clint says who the fuck chased him in an aggressive tone. The senior of the group goes up to Clint and started this stupid monologue about him being O.G (Original Gansta), and a few minutes into it Clint kicked him in the balls, pulled his shirt over his head, and delivered upper cuts.

This knocked him to the floor. We told his friends to stay back or you're next. When crazy Rudy started kicking him in the back with steel toe boots, they just stood there. I felt sorry for the guy after the fifth kick and I saw a helicopter shining its light, so

I grabbed Rudy,pointed at it, and said "let's go, the hoodacopter is coming," which was our slang for the police helicopter, so we left. Another time we were walking to the store. It was me, Anthony, which we called Ant Rock because he liked crack, Joe, and Gabriel. This guy almost hit Joe while pulling into his driveway, and as usual, we yelled at him. Instead of just telling us to piss off, he jumped the fence, got into Anthony's face, and they started fighting. Anthony started losing, so he said get this mother fucker. Immediately we started pummeling this guy into submission. When everyone started running he grabbed my foot and did an ankle lock. I couldn't go anywhere, so I kept uppercutting until my hand was covered in blood. Then he says "look who's going to jail," and I scream "get this fucker off me," and one of them came and stabbed him three times in the back. Bubbles and blood oozed out and he finally let me go.

To get away, we ran through allies and small roads to get away, avoiding the main streets. We went to Joe's and avoided that side of the hood for a while. We never fought each other, except for the one time Willis was talking about his girl, who is now his wife. I said "fuck that bitch," all drunk, and next thing you know I am on the floor and my head is bouncing between his fist and the street. When I realized I was in a fight, I threw three or four massive blows and it was done. I went to grab a beer.

I honestly didn't care. We were friends and nothing could change that. This guy Chris would come over after he broke up with his girl Alicia, which I got with a bit. One day he pulled up in this stolen van with two girls in it and said "hop in homie, let's go party." I jumped in and was ready to party but that did not happen.

As we pulled off I sat right between the two girls and put my arms around both of them, which made Chris a bit jealous or curious. I say that because he drove with one hand on the wheel and kept an eye on us instead of the road. This is why he crashed into the next door neighbor's car, who was a cop, scratching the whole left side of his Mustang and the right side of the van. Chris and the girls bailed, so I did too. As for the van, it coasted down Earll Drive until it crashed into the light pole at the end. I don't know where they went, but I ran to the next street and came around the other side like nothing ever happened.

The cops came and I was just chilling, watching them file the report. The things we did at his house were not always thuggish. There was some childish fun as well: for example, the huge rubber band fights we had. One day we had this huge bag of them for some reason. They were in many different sizes, from thick, thin, tiny to huge, so we stuffed our pockets full and agreed to start the free-for-all war. Five minutes after breaking to give the seven of us time to set up whatever spot we chose to set up at. Five minutes went by as I waited at my position in the bushes, and my first target, Gabriel, came creeping by and didn't see me. I let it rip with a thick one, hitting him square in the back and I could hear it hit like a slap. He yelled "ahhh" and I took off to find another spot. I came through the door and was surprised to see nobody in the living room. I headed for the hall, but when I turned the corner I got hit in the chest hard by two thick ones, tied together by Joe's brother Mikey. I jumped out of the way to the kitchen and hit Clint with two shots because he missed. He ran so I went to the back and got little Johnny, the youngest brother, and Sarah, his little sister. Joe's cousin Matt darted across the yard and I got him in the back.

I felt like Solid Snake from Metal Gear Solid. I decided it would be a good idea to position myself on the roof, which worked pretty well, since I was picking everyone off at an alarming rate. I stayed there too long I guess, because I already gave my position away. Here comes Joe. I didn't hear him till he was about 15 feet away, and I turned and immediately fired band after band. All these shots did not matter. Although he had a pocket full, he had a frying pan blocking every shot as he got closer. Soon he was right in front of me and smacked me upside the head, sending me rolling down the roof, hitting the ground. I thought about trying it again.

It was like a ride, I thought as I popped right up and kept firing at him, eventually getting him over and over.

That was one of many rubber band wars we had. We just had to collect all the bands after we were done, and we used them so much they all broke because they lost stretch and got dry. After they all broke, there were no more wars. I can't even remember where the huge bag of rubber bands came from, and neither can Joe, but it was really fun while the bands lasted. They lasted a very long time too. One time Gabriel was running to catch a football out front and fell on to a wicker chair. His arm somehow got tangled in it while rolling and broke. When he got up, it was in the shape of a "u." I was surprised he didn't pass out. The sight of seeing a NFL player hyperextending his knee made him almost pass out before.

Chapter 5

Getting Into the Dope Game

Gabriel got interested in how much money Cali Johnny, as we called him, was making. He was considering trying to sell weed himself. He didn't have the money to start and Johnny was not going to front a novice seller, but somehow got this guy outside of our hood to front him a quarter pound. I went with him to pick it up and this girl was there, so I talked to her and got her number. That night as we chilled, I listened to "Who Do You Believe In" by Tupac word for word. I was technically atheist, but after really listening to it, I decided right then and there to believe and became agnostic. God works in mysterious ways. Einstein once said "coincidence is God's way of remaining anonymous," so when you see coincidence, keep that in mind.

My brother started smoking it more than he was selling it, and I figured he would get better and I told him to sell more, smoke less. After a few weeks, the guy wanted his money but he didn't have it. Then, a few more weeks went by and the threats started. That was not good because all that did was start us conspiring on how to get him before he did anything. I thought, "you know what, I can fix this." I told them to chill and I arranged for me to meet him since the girl there was his sister. I convinced him to front me a quarter pound, and within a week I was back for two. Before you knew it I was buying pounds at a time within a month.

I was 16, starting to make good money. I simply had a product and all I needed was clients. I would sell to the homies as I called them, but I also would go to parties with dimes, nickels and quarters. Those are different measurements of weed. I would take out a joint and smoke it, which attracted buyers. It was easy and sometimes I would hook up at these parties. One time me and this girl had sex just outside the party and came back to it. I was sure to practice safe sex apparently, since I woke with the broken condom still on. I woke up on the couch at this girl Helen's house one time. She was beautiful and lived two houses away from Joe. After a while the parties were too much for his mom. I guess they were becoming bigger and bigger. Although we did make sure to clean up, she did have a new man, so that may have had something to do with it.

Chapter 6

Selling Weed aka Dope

Cheech and Chong may call it dope, but I call it weed. I didn't think I would be that good at it right from the start, but there was always room for improvement, I thought. We did not hang out at Joe's house any more, we just stayed at the house, and with my new phone I was getting even more clients. At one point, staying home meant getting bored, which lead to us doing something stupid and somebody getting hurt. For example, at my nana's I was kicking a beach ball until I accidentally did half a backflip and landed on my chin trying to kick it too hard. I ran to the bathroom, looked in the mirror, and it looked like I had two mouths. It was so bad. I held the flap of skin up, let go, and it just fell open.

At my mom's horse playing, I pushed Gabriel almost all the way through the sliding glass door, cutting his brachial artery. Many other incidents that happened, such as a nail through my heel running barefoot in the gravel and more. My mom ended up dating this bartender from Six Shooters Bar and Grill, which is now Desert Sky Mall. Dave suggested that I work at the restaurant as a bar back. I thought about it for a while and saw it as an opportunity to get new consumers. I was a customer looking for consumers to make more money. Business is something I have always understood for some reason, so I took the job as a bar back. I started giving out my number to people that smoked and I could usually spot them. Within no time I had way more to sell to, and if I was at work my brother handled it.

A bar back has many duties, such as connecting and disconnecting the kegs and soda in back, as well as restocking the liqueur and garnishes like lemons and cherries. My favorite perk was being there for the wet t-shirt contests. Gabe was just a call away. As money kept coming in, I decided I needed a bank account, so I opened one at Desert Schools Federal Credit Union. No more than $9,000 deposits until it was all in. As time went on, the money kept rolling in. We just had to follow a few rules, like not selling to people you don't recognize, and keep it pre-wrapped and fresh.

I didn't like going outside the neighborhood to resupply, so I found some suppliers closer. We will call them Chico and Big Boy. Chico was a big Mexican guy and Big Boy was a big black guy. I would get four pounds at a time from one or the other, wrap and seal each pound separately, using a pound at a time, selling no more than a half ounce. Doing this kept it fresh and customers came back, turning each $600 pound into $1200 easily. This went on without my mom knowing at all. She never asked about all the new stuff like the PlayStation with all the games, stereo system, and much more. Like my movie collection, which started with five movies and kept growing very fast, included The Last Dragon Unbreakable, I'm Gonna Git You Sucka, Higher Learning and 1984 since we read and watched it in elementary. The Last Dragon is my number one favorite because it's a love story, but my new one is Free Guy, also a love story.

We also went to California and Catalina Island, and rooms in Catalina cost an extra $100 since we were underage. The activities we got into were parasailing, banana boat rides, submarine tours and something with the dolphins. We would visit Vince, my biological father, but it just showed me how unlike him I am, and how I don't want to be. Visiting him ended the day we got into an argument over me smelling like weed and being high. We could have bought a car, but getting places was never a problem for us, and then would have insurance to keep paying which is why I got rid of my truck I had now. A few times I got burned by people using fake money, so I got a black light in my room and glued glowing stars on the ceiling. Fake money glows differently. These people never came back around, of course. They knew what they did. Fake money glows very bright. I even kept the fake money as a souvenir and a reminder to trust no one. The only ones I trust are my friends since childhood, aka home boys and brothers-in-arms.

I had a phone now, so it was easy to reach me. I no longer had to go to parties and what they called kickbacks to look for my consumers. They came to me and my memory was really good, so keeping track of who was who was an easy routine despite me being really high most of the time. Standing on the street corner like the crack dealers did not appeal to me. That's like clocking in and putting more work in than payment. I was always on call and on the clock from home as long as there was money to get, and there was plenty of money to get. School was still a priority because I wanted that diploma, which I knew would open many possibilities.

I was eating cereal and watching the news when September 11th 2001 happened, I watched both planes hit and I remember thinking how could this be happening. I was shocked. I felt sorrow for my fellow Americans, I was no patriot but I grew up at a time when we said the Pledge of Allegiance in class and I felt united. I felt bad and stayed home that day to come to grips with what happened. As I watched I found out more planes had crashed and it almost seemed unreal. Though I sold from sixteen to nineteen and made a bit of money, it is hard finding much to say about it. It was pretty uncomplicated and basically cut and dry. The more clientele you have, the more product you need, and you need a supply. Then presto, you're making money.

Chapter 7

Danger of Selling

There are many dangers to selling, and they all come from people. For starters, cops are definitely something to look out for. You have to be covert when you sell. I did this well, but when we went to Lake Pleasant one day, I figured I wasn't selling, I would just take some and keep in my tent to smoke. That was stupid, because the cops searched my tent and found it. This earned me a felony bumped down to a misdemeanor, thank God. That would have stopped me from joining the Army. Another time that the cops almost got me was when my tio Ric- uncle is tio in Spanish- needed some weed. I brought plenty but all he needed was a little.

My tio Mark was there this time. He was the first bully in my life. He used to take my candy and shove me around. He tried shoving me around, and with one punch I shattered his zygomatic bone, which holds your eye in. His eye immediately started falling out, so he quickly held it in with his hand, and I was both a little shocked and grossed out. I left just as the cops turned on my nana's street, and I passed by on the bike with an ounce on the way home. A close call. It would have been worse if I stayed, as they would have gotten me for both assault and the weed.

There are people with fake money also, which is why I got the black light- fake money glows. I got guns just in case, and carried one around the hood sometimes because we did get shot at sometimes: a few times at Starlight, a few at El Oso Park, and a few in the neighborhood. I had a 45 caliber Ruger, steel Mossberg pistol grip shotgun, and a double action 22 caliber bobcat. My favorite one is the bobcat because double action made it very fast, and it has gold plated trim, which I still have to this day. We had to worry about being robbed at these parties and came close once, but not when you're operating from home. We had it made. I would walk around home so high, wrapped in a blanket while my brother got really into his Sega games.

As I walked around the kitchen wrapped in my blanket and munching on Skittles, Gabe was in the family room, which is accessed from the side door, and was unlocked for some reason. Three guys come in through the unlocked door with pistols and say "give us the fucking money!" My brother, who happens to be playing Beavis and Butt-Head: The

Game, says like Beavis, "are you threatening me?" then snaps out of it. When he did he looked at me and said to get the shotgun, I was gone in my room before the blanket hit the ground and they ran. I came back and they were gone. We just laughed it off. We were obviously too stoned to pursue, but it was good with that shotgun, and we were damn near surgical with it. After that incident, I started thinking about the end result of this choice of lifestyle. There is no health care or retirement. It was illegal, so there was no calling the cops for getting robbed.

At some point we would get too big to be discrete. We tried, but it turns out, now that I talk to all these neighbors to this day, they all tell me they knew. I suppose because I kept things quiet they didn't bother me. Big dealers like Chico and others ended up getting too big and eventually got raided, arrested, and property taken, except the house which was in his mom's name. I continued with the usual daily routine of playing my games, watching movies and serving the consumers. They called, I got it ready, and when they knocked it took one or two minutes tops to check the money exchange and they were gone. I didn't have time to talk since most of the time I had my movie or game paused. Friends would come over also and we would pay a bum five bucks to buy our four bottles of Allize with the money I gave him. We didn't need to steal and we never begged like people do now. Matt, Joe's cousin, Joe,

Gabe, and I all had a bottle and a big pile of weed in the center of the table with some 2Pac playing. All of this was great, but that lingering thought of where is this going and what is the end game just bugged me more and more.

One night I was playing some Bruce Lee game on my new Xbox. I just kept buying movies, games and systems. Dave came into the living room to watch me play. After 15 or 20 minutes he started talking to me about joining the U.S Army. He started telling me all the perks and benefits. Itall went in one ear and out the other, except for change of scenery and Montgomery Government Issue Bill (MGIB) for college. That night, as I looked at the stars all over the ceiling, I really thought hard about my future and the direction my life was headed. I often asked my mom why these people live on the street and do drugs. She simply said their head is not right and has no long term goals. Other than their daily goal of getting high, as well as other short-term goals, they made no progress in life.

I decided at that point I needed a long-term goal as well as a change of scenery, so the next day I went to the recruiting office with Dave. I walked in and stood before the recruiter, who looked a bit like 2Pac, and said I wanted to join. After doing all the paperwork and setting up my swear-in date, I entered the Delayed Entry Program (DEP).This meant that I was all signed up, and my report date for basic was set for Ft. Leonard Wood in Missouri for One Station Unit Training (OSUT) on March 10th 2003. Though my report date was six months away and I was all signed up, I still had a business to attend to and I planned on doing just that until it was time to go to basic and AIT.

The MOS (Military Occupational Specialty) I chose was 12B Combat Engineer. The way it was chosen was funny. He said, "did you like playing with fireworks as a kid?" I said "hell yes" and he showed me what they do. Funny thing is, the MRE (Meal Ready to Eat) sealed the deal, actually. WIC is a state nutrition program for women, infants, and children that my mom often used, and I loved the government cheese they gave us. The chicken noodle had liquid government cheese in it!!! Between my recruiter looking like 2Pac, the MOS, and the cheese, I was sold and couldn't wait to sign. The date for my report to basic drew closer, and I didn't do less stupid things but I should have. I was running through the street drunk, firing my 45 in the air and doing beer runs for no reason until one night, around Halloween, I tried messing with decorations in somebody's yard. I said look, I'm going to dance with the skeleton. As I pulled, I realized it was tied to the ground. Next thing I know, I'm tackled and the cops showed up. I said "you don't want to do this," and they decided to let me go, thank God.

Chapter 8

Joining the Army

Until the talk with Dave and almost getting robbed, I never really thought about the future. I just looked a year at a time into the future. I became curious about what I would be doing five or even ten years from then. That was a big factor in my decision to join. Time went by quick and it was March 10th 2003, my report date, and I was 18. We were watching the initial bombing of Iraq before they invaded. Suddenly, there was a loud knock at the door. It was my recruiter here to take me to the airport. I gave the family hugs and headed to the door, and the look on their faces was as if I was going there immediately. I took care of supply for Gabe and bought four pounds for him to sell a week before.

Gabriel and I flew often, so flying and traveling to Ft. Leonard Wood, Missouri was nothing new to me until I got to processing. All of my personal belongings had to be put in a black plastic bag, and I was issued all of my uniforms and gear. We were only allowed to wear our PT (Physical Training) uniform and running shoes issued to us during this process. This process involved paperwork, shaving the males heads, collecting height and weight information, but the last thing was painful. We lined up and walked between two lines of people holding needles. We walked through getting shots in our arms from left and right. Ouch.

I thought it was over, but I got one more in the butt in the next room. The doctor said to me, "they call this the peanut butter shot" then stuck me. I soon found out why: it felt like I had peanut butter in my right cheek. All of this processing took a few days, and afterwards we just had to wait for a unit to be ready. OSUT meant the unit would be our basic, and AIT (Advanced Individual Training). Time went by and for some reason this guy, Ford, had a problem with me and wanted to fight me. I just turned a new leaf, so he was lucky. I would have, but I had to prove to myself that I could do this army thing.

The day finally came when I was told I was going to foxtrot 35th OSUT, and I was excited when I should have been nervous. I couldn't help it. I had no fear, like the American hero Chris Kyle, the American Sniper. The harder it got, the more I seemed to love it. I didn't prepare for this at all, but approached the situation as if I did, and when we got there a rude awakening was waiting. We got in the cattle truck with a rucksack duffel bag and trash bag of personal belongings, and we rolled out to F35th. When we got there, ten drill sergeants- two female and eight male, five black, four white, and one Mexican guy got in our face yelling at the top of their lungs. Move with a purpose private, go this way, go that way, do this, do that. This flustered and confused most, but not me. I just smiled. They were definitely intimidating though, especially Senior Drill Sergeant Davis. He was big, like Micheal Clarke Duncan from *The Green Mile*.

After all the yelling, having us lay out our gear and doing some push ups, I was assigned to 2nd platoon and my bunk. Each room had about ten people in it, and out of three floors I was on the second. They briefed us on the schedule for tomorrow and told us to square away our rooms. Day one they woke us up by banging on a trash can while walking down the hall. As we formed up, we were told that our first physical training test will be done. We start with a PT card recording height, weight, name, and age. The first events were push ups and sit ups, which were recorded on the card. The two mile run was the last event, and although I ran, it was never this far. I was in for a reality check. I ran those two miles and I felt like my lungs were bleeding. I swear I could taste it.

Basic training was learning drill and ceremony, PT formation, repelling, rifle marksmanship, and other basic skills, as well as that obstacle course we did often. I got really good at crawling on top of the rope, which was called commando crawl.

Chow time in basic was very strict. You got a tray and side stepped as they put food on your plate until the end, and went to the table. We had to fill table after table without picking our seat. After sitting, they said we don't have time to taste it, eat it and move out. Those that took time became the reason the ones outside formed up and got smoked. Being smoked is basically doing PT constantly, and by this time we are in BDU (Battle Dress Uniform), and they get very hot.

We went back to the barracks and they trashed everyone's room, then showed us how the beds are made and how every room and closet should look. Now that we knew the standards, there was no excuse, and was day one showing us the standards as well as duties. Dinner chow was at 5:30, 1730 military time, and by 6pm we were at the barracks. At 8pm the drill sergeants yelled for us to get in the hall where they showed us how to toe the line, that is basically lining up shoulder to shoulder so they could inspect us. Face shaved and nails clipped was the standard. Flip flops everywhere from the shower to the room was also the standard. They yell, prepare to post, you get next to your bed, then bunks post, and you're in bed. Time went on and I got way better at running, probably because I have Tarahumara in my blood line, it's a tribe also called the running people. I didn't know about it until I was 30.

After the PT run one day, I thought I would skip the shower and get dressed and it was not good. Drill Sergeant Akers was this black guy built like Tommy Herns. He said to me "why didn't you shower, private?" I said "because I didn't sweat, drill sergeant." Bad idea for me because as soon as I said that, he said everybody put PT's on and go to the war room. The war room is a big open bay. We all crammed into it, and as soon as we all did, it was on: front leaning rest position, move, push ups, basically in cadence exercise one two three one two three. We moved to jumping jacks and other exercises until the wall was sweating and the windows fogged up. Then Drill Sergeant Akers said to me, "are you sweating now?" I said "roger, drill sergeant" and he stopped smoking us and sent us to the shower. I learned my lesson and people got a little mad at me.

I had a bit of a bully problem for a moment. This big red head guy in our room kept running his mouth saying child-like insults to me. I got tired of it and chucked my boot at his head, then he rushed me and I guillotined him. He kept his mouth shut after that. Another time they had us duck walking up and down the halls. That was bad, but crab walking was worse, which we did also. We got smoked like that often, sometimes for no reason at all. San Juan Hill, they called it, was a hill next to the barracks they smoked us on so much. I loved it. I would be smiling while everyone was emotionally distraught. That was learning the ropes of basic military life, like drill and ceremony, general orders, marksmanship and getting smoked. Breaking you down so they could build you up, they told us. It is just what they did and we thought it was for no reason. When basic was done, I was so much stronger and my attention to detail was amazing. Next was AIT and I couldn't wait.

AIT was way different from the PT test. There was no yelling at us constantly, we just received our card. We filled out day one. Once we got it we went outside to formation and began. After push ups and sit ups, we did the run and I flew through like nothing, getting 12 minutes. My lungs didn't feel all messed up and I did not taste blood, thank god. That scared me a bit. I was even more shocked at chow. We formed up and they dismissed us after saying what time to form up after chow. We were given way more freedom.

I had no religion but believed in God, so when someone needed a buddy for church I went. This introduced me to many different religions; there were three or four in our barracks. First was Christian, then Catholic, Greek Orthodox, Mormon and Methodist, I think. Combat engineers deal with explosives, so we were given demo cards as we began learning MDI (Modern Demolition Initiators) and knots. Knots are important because they have many functions and uses as rope and demo. I like the prusik because you can use two to climb rope. The demo card is a reference because it is very much involved with demolition. Many formulas all have to do with math but not all. M-1 dynamite, C-4, shock tube and det cord is what we learned, along with the demo card of course. Once we learned the basics of MDI, knots, and explosives, we finished by learning how to prime as well as detonate a shape charge and crater charge. Last but not least, the Bangalore. Only things to learn for a basic combat engineer were obstacles: obstacles such as mine fields and wire obstacles. Combat engineers specialize in basic mobility and counter mobility.

A week before graduation we got a day pass and I went to the store for some junk food and hung out at the rec center. The day before our ceremony they read and gave us our orders. My name was called so I got up and got my orders. He said I was going to Camp Howze Korea. All he said was "Soju is the devil, be careful, private."

Chapter 9

Korea

The graduation ceremony is very hazy in my memory. I do remember them playing the song "Fighter" by Christina Aguilera. We did the whole thing wearing our class A uniforms for the first time. I felt like I accomplished a very life-changing feat. I definitely was changing the course of my life. I didn't know how massive of a change it was, but I was about to find out. We returned to the hotel afterwards and caught the next flight from the airport home. I went home with my TA-50, which is the duffle bag, and a ruck stuffed with all my standard military gear and personal property. I didn't even change out of my class A uniform.

In no time we got home, and then I got out of them and smoked a joint. I was smoking a joint because I just completed a hell of a long-term goal and I was told they don't piss test you in Korea. I was also told that they don't deploy from Korea and have not since the 60's. The fact that I would not go to Iraq was disappointing, but on the bright side I can smoke my joint, and I was so high. This was around June and my report date wasn't until late September, so I had time to celebrate and be happy. I didn't check on how business was going with Gabe because I was going another direction in life. I had nothing to do with it. I made enough money and I was going to Korea as a U.S Army soldier.

My birthday is August 22nd and after a few weeks I was anxious, so I flew out and reported to camp Howze Korea early at age 19. I remember thinking, wow, I am 19 and going to Korea with a five figure bank account. The plane took me to Chicago O'Hare Airport, where I got on this huge plane with "Korean Air" on the side of it. I checked all my TA-50 and suitcase in Phoenix, so I just had to jump on. I sat down and the stewardess asked me if I wanted a drink. To my surprise, wine was mentioned. I am not passing this opportunity up, I thought to myself. I said "a glass of wine please." The glass of wine was tart but it was my first glass of wine, so I had to finish it. Turns out the drinking age in Korea is 19, and the day you're born counts you as one, so technically I was 20 there.

The flight was ridiculously long, I thought to myself as I drank another glass of wine. Thirteen hours of flying and I'm at Incheon Airport in South Korea. I arrived during the whole bird flu pandemonium, so most of them had masks on, looking foolish. I took a shuttle to Camp Howze and found the Higher Headquarters Detachment (HHD). Once I reported with my orders, they took me to the CIF (Central Issue Facility) to get extreme cold weather gear since it gets at least 15 below. Winter was only three months away, so I would need it for sure.

This company was not what I was expecting. I was issued my room and told once a month or so we had alerts. During these alerts, we had to have everything packed and ready for combat, but this company didn't bother and just had soldiers report to the company. We did not go to the field like we did in basic. It was more like staying at a hotel and playing video games. Setting up the big green tents and tearing them down was our only duty, and cleaning them then returning to our cubicle mobile-home-like room. Not the soldiering I imagined myself doing. I did not enjoy feeling like a pogue. That is military slang for non infantry MOS personnel. You could say I was about to learn a lot of slang and acronyms.

Since I wasn't soldiering like I was looking forward to, I got into hanging with a group of soldiers. They just hung in the recreation room and played drinking games. Thompson, Walton and Schwartz taught me how to play Captain Dick Head. All 52 cards face down and spread out is how you start. Red you give and black you take. For instance, flip a ten red and you can have a player take all ten, or you can split it, and black you must take ten drinks. Jack is a waterfall. You start drinking when the one on the left begins and can't stop until they stop. Queen is rhythm and rhyme, so start with any word and the one that can't rhyme the next word eventually has to drink. Car, bar, star for example. The king is association, so for example, types of cars or whatever until somebody gets stumped then they drink. Ace is Captain Dick Head, and two or three are major and general. Also you get to make up any crazy rule each ace you draw. I like the invisible man. It always gets them. If you don't take the invisible man off your beer before drinking, you double your drinks.

We didn't go to the ville, a nickname we called the town beyond the gate at night. We just played drinking games. One night we played the game. Thompson was losing badly and I was captain within my second turn, and major by my sixth. Invisible man was what I chose first, of course. He was drunk and kept forgetting, doubling his drinks over and over. Finally the waterfall came and I started. By this time he is not looking too good, I remember thinking. He started dry heaving, so I pushed my chair back away from the table and thought, oh shit he is going to puke. He put his hand over his mouth, trying to hold it in, as his mouth filled and puke sprayed from between his fingers in many directions. With his eyes wide open and one free hand, he ran for the door, put his hand on the door knob as he looked at us and opened the door to puke outside.

The recreation room was a mess but we managed to clean it up. The tours I signed up for were so much fun. We got on a bus with plenty of OB (Oriental Beer), Korea's most popular beer made there. We did karaoke as we made our way to a nice restaurant, starting with Hotel California. This is when I discovered karaoke is awesome. We got there and waiting for us was a huge feast, there were Korean main courses for each of us, and appetizers lined up in the center of the table at arms distance. Bulgogi and rice, which is beef and rice, yaki mandu, kimchi, glass noodles they call japchae, and chicken wings to name a few. I didn't last long at this HHD company because soju is the devil. The drill sergeant warned me about this but I didn't listen. Korea sells soju in 12 ounce bottles and these tall plastic gallon bottles.

We had the big bottle drinking in Walton's room, and at one point I remember chugging the bottle as they chanted "chug, chug" and I blacked out. All of a sudden I'm standing in PT formation all dressed holding one leg back doing the thigh stretch. I was still drunk so I fell on my face. I got up stumbling and the SGM (Sergeant Major) noticed. He ran me around the gym until I puked, told me to drink water, and did it again. After we went to the company for a breathalyzer, and I still blew a .08% BAC, the commander said it was Chapter 13 discharge or send me to a line unit. It was decided that I go to a line unit, so I went to Second Infantry Division 2nd ID 44th Engineer Battalion, the Broken Heart Battalion, which they said I would hate but I loved it. The Broken Heart Battalion got the name because they were just finishing repairing bases damaged from WW2 after they arrived in 1948. They came home for a short time and went to the Korean war. Leaving a trail of broken hearts, Anniston newspaper of Alabama wrote. I was going to be soldiering like I hoped.

I was assigned to Bravo company 2nd platoon with Staff Sergeant Barber as my squad leader. I got my room key and moved into the second floor of Castle Gray Skull. They named it that due to the long staircase leading to it. Standing at the company, that's what it looked like, I was made a M113 driver, which I did not like but eventually got out of. I messed up constantly on purpose, finally wrecking a water buffalo and getting fired. My first encounter with my squad was ok besides the fact SSG (Staff Sergeant) Barber was calling me a spic, trying to get a rise out of me. Guess he saw the thug in me wanted to test me. I found out he wasn't racist, he was just a hard ass.

The rest of my squad was Sgt. (Sergeant) Glick, my team leader and Sgt. Roberts, the other TL

(Team Leader), Steele, Conte, Szcypka, Henderson, and Gaona. First night we had an alert at 0100 (1am). I had my gear packed and ready according to the packing list, I grabbed my duffle bag and ruck out of the wall locker as I made my way to the motor pool. I prepped the vehicle through gear where it needed to go, and got in the proper spot according to the op order. Once fired as a driver, it was a tiny bit easier, but I still helped Conte prep his M113. It was October and I went to the Ville for the first time with Steele and Szczypka, nicknamed Z. Three KATUSA (Korean Augmentation of The U.S Army) as well as Moon, my roommate Sgt. Lee and Park. We were in for a culture shock. Many Korean people wore masks due to the bird flu, but South Korea is a bit communist and highly influenced by the media. Every military age male must serve two years.

The shops, restaurants and bars were awesome, like the shop that chips your PlayStation so you can play the bootleg games they copy from the real ones. It was three for $10. Mom's sold chicken and cheese Ramen, yaki mandu, and other things. The Filipino restaurant had Torcino, Bulalo, Lumpia and more. It was next to the phone shop and liquor store. Vendors came out at night with chicken on a stick, and the clubs opened at seven. So many clubs, about 11 or more, had strange names: Pop's, Elvis Rainbow, and Smile, to name a few. It was like a game finding them, going down a new alley and stumbling on a new one. There were what we called "juicy girls" or "drinky girls," because they sat and chilled with you if you bought them a drink, which is just juice. "Buy me drinky," they would say. I would have three at once in the booth just for the hell of it. There were Russian, Korean, Philippine girls and a girl from Thailand. I would also sit with my peeps and buy a soju kettle, which is Champagne, soju, OJ and 7UP. I would wake up not remembering how I got to my room in Castle Gray Skull often. We partied hard but we also trained hard. As long as you did your PT and didn't fall out they couldn't care less. Marta, for example, was puking during the BN (Battalion) run lTC Mize was leading, but he finished it.

November approached as we prepared for a two month field problem where I tore up the water buffalo. We got out there in the -9 degree weather, set up camp and communications, and by night we finished. The point was to practice individual tasks, squad, platoon and company tasks, but the way we communicated was on point. With Sitrep's radio checks, nine line medevac and just overall situational awareness, I really felt we got better each time. This was my first time, so I pulled the radio guard first and we made sleeping arrangements inside the track. I didn't like the idea of that, so I thought it would be better to sleep in the trailer. I could find my bag, I thought, and get a cold weather sleeping bag. I had no light because the flashlight was packed in the locked duffle bag, so I did my best to wedge myself between the bags. Once wedged in, I curled into a ball, put my hands under my armpits, and prepared for the worst night's sleep ever in below freezing temperatures. I know it was below because when I emerged to pee, the puddles were frozen in the morning. I never was unprepared after that lesson learned. That taught me to think ahead and plan ahead, lay it all out, plan in your head and even on paper.

We came back to Camp Howze from the training grounds at the beginning of January. We went to the field for two months at a time. The base had a barber shop, dry cleaning, and many little noodle shops that seemed to be hidden on base. Right after the front gate we called "the bottom of the hill" was Popeye's Chicken, Anthony's Pizza and Subway. I fell in love with Popeye's in Korea. Popeye's Chicken is the shiznit, I thought while watching my bootleg Little Nicky movie. I often got five movies for ten in the Ville. Last but not least, the free movie theater was where I first watched *Tupac Resurrection* and *Passion of The Christ*. 44th Headquarters was at base and Bravo at the very top of the hill. Then there is the barracks.

Bravo was a very good unit but we didn't have a commander that was serious. Cpt. (Captain) Ho loosely commanded and we got by because we took initiative. "Change socks and drink water," he would say so much it became his trademark. Our First Sergeant I can't remember, but soon we got Cpt. Blackmon, a ranger, and First Sergeant Douglass Shepherd, a Sapper. A New LTC (Lieutenant Colonel) Tommy R. Mize came soon after. Charlie's company had First Sergeant John Etter and CPT Czekanski. I can't remember Alpha. I think they had Cpt. Law. Every club was playing Yin Yang Twins, Tupac, Outkast and more, but within 15-20 minutes you always heard a 50 Cent song in any club. It sounded like he may have cashed out like I did, I thought as I listened to the lyrics, but I shrugged it off and just had fun.

Steele Henderson and I went to the red-light district a few times and it was interesting. We often went to this place called Gecko's where the girls spoke really good English, and there were no "drinky girls." Good times. I learned so many Hangul phrases from my roommate. It wasn't a problem. l learned some Russian and Filipino too. We went back to the field under the command of LTC Mize for one more cold weather field problem. I was made grapple man for many of the breaching missions we did. SGT Barber looks at me and says "LTC Mize is out there watching this next obstacle, go kill it." That's exactly what I did, running out like I was storming the beaches of Normandy. I ran in three to five second rushes, throwing the hook, pulling it in, and repeating till the Bangalore team had a clear path to the obstacle. Sounds easy, right? It was very exhausting, especially when you're doing it in cold mud.

After the field, we came back, washed our gear and vehicles at the wash rack, did PMCS, and laid out gear to inventory. I loved the harsh training so much that I opted to stay another year, which you get a $2,000 bonus for. I got my bonus, did my leave papers for August 1st to the 23rd, and booked my flight three months ahead of time. Meanwhile, I just kept having fun. We had barracks parties too. Nobody did it like Bravo. Imagine everyone's fridge in the barracks on all three floors full with OB or Bud Ice. I left my door open and yelled, "if you need a beer, grab one," as I left to explore the barracks. You could walk into any room and hang out or hang in the halls, and even the KATUSA's drank a bit. Summer was here, so our barracks parties were a must. During the day, it was so humid you would sweat just walking down from the barracks to the company though.

My time to go on leave was finally here, and I took a taxi to Incheon Airport and did my 13 hour stretch to Chicago O'hare. I went from Chicago to Phoenix with my duffle bag since I had no suit case. Once I got home I asked Gabriel how business was going, and he told me he had nothing: no money and no weed. I just laughed, shook my head, and said "wow I'm gonna get an ounce." I honestly didn't care, I was here to celebrate my 20th birthday. I invited a bunch of people and made a soju kettle with the big soju bottle I brought. Everyone was buzzed pretty good. I started smoking weed and didn't stop until it was time to go. I spent the whole month celebrating my birthday, then it was time to go. I took the same route back to my BN and I signed in around August 23. The staff duty sergeant said "you're just in time. We are deploying to Iraq."

I headed to my company to check in and was told our brigade and battalion had orders to go to Ramadi, Iraq Al Anbar province. I got dressed in uniform and went back to hang with the platoon in front. As we sat there waiting for our leaders to plan during the meeting, this soldier started talking bad about our country in general. I think fear had something to with it and I remember thinking I know I stand with my country and its people. I definitely hoped I didn't have to depend on him in combat. Unruh showed up and shut him up. Robert Oliver Unruh gained hell of respect from me that day. I got up, fist-bumped him, and said that's what I'm talking about. Every time we crossed paths I'd nod and say what's up battle and he would and say what's up brother.

The next day they were already packing up personal items and loading them into a connex, which is a metal shipping container. I used my phone card and phone to call my mom and tell her what happened. We palletized our gear, prepped our vehicles to be shipped to Kuwait, and had to wait till our short train up. By this time we had been through plenty of random alerts and field problems. It was easy to shoot, move and communicate so to speak. We had an alert but we were preparing to leave and fight, so we just changed our routine a bit. Everything went smoothly and it was done quickly. We had new DCU (Desert Camouflage Uniform) gear and RFI (Rapid Field Issue) gear. All there was left was our short train up.

Steele knocked on my door and he had a case of OB. You want to watch Passion of the Christ, it's at the theater he says all up beat. I said hell yes, let's get it, so we made our way to the theater. The theater was free and we could bring beer in, so we sat there watching and sipping away. This was a very moving movie based on real events. Although the actor for Jesus was white, I'm pretty sure he was most likely Arabic and looked like the people in the area of Ramadi. Ramadi is 5 hours away from Golgotha, Iraq, where this all took place, which I didn't know. I thought about the Iraq location and the story of Iraq people. He was probably Arabic. During the movie he says we should all go out to Ilsan and have fun before we start training. He meant me, Gaona, Henderson, Z and him. So the next day we got a taxi and went to Ilsan. Bali Bali ahjussi, I said, which means hurry hurry man and they go really fast. It's scary and fun.

First Sergeant Shepard gave us an overnight pass because at midnight everyone had to be on base. We get to Ilsan at about 1630 (4:30pm) and go find a nice place to eat. I can't remember the name but it was some good food. Then we went to the clubs, in each club there were girls, but not "drinky girls." We talked and drank with a few, and moved on to the next, using phases my roommate Moon taught me. I was doing good. We all had a lot of fun going from club to club, drinking and chatting, but it was getting late and we needed somewhere to sleep. I look up and see a big hotel from where we sat. I will call it the Butterfly Hotel. I say, let's go stay there, so we walk our buzzed selves over and get a room. Nothing out of the ordinary, so we get in the elevator and go up.

Once at our floor, we all get off except Steele and Z. They head to the top floor. We find the room and as we are settling in, Steele and Z come bursting in screaming. "Holy shit, holy shit what the fuck," Steele says. Z says "fuckin a, right doggie," cause he loves saying it. Finally, after calming down, they told us that there is a brothel on the top floor, so of course we had to check it out. We got there, paid our fee, and waited in the waiting room, watching T.V. as they brought us six packs. One by one they took us back. Finally it was my turn. I went to the desk and the attendant must have tapped a button. The bookshelf next to her opened like a door and out came a girl dressed as a maid to escort me and my sixer to the room. She brought me in and scrubbed me down in the shower room from head to toe. Then she brought me to the massage table. I got a really good massage. She walked on my back too, popping all kinds of stuff. We eventually ended up in bed twice. I woke in the hotel room to Z yelling "what the fuck am I doing in the bathtub." I see him get out confused.

We got our stuff together and headed back to Camp Howze. We found ourselves watching combat movies like Saving Private Ryan, Black Hawk Down, and others. Trying to understand what we may expect, I guess. We soon went to train up and it was short. We worked on urban combat at a MOUT (Movement Over Urban Terrain) site. Afterwards we palatalized our equipment, prepped our vehicles to be shipped, and got on buses to Osan Air Base. I remember sitting on the bus wearing these brand new DCU's and my weapon, thinking about how surreal this is getting. As I looked out the window I could see the plane. We filed out and got on the plane. Apparently, our BN and other units were the first to deploy from Korea in 50 years, which is historic. This is an elite unit. I'm good to go.

Chapter 10

Ramadi, Iraq

We arrived in Kuwait about 1630 where we had to stay on base and climatize a few days. We had to wait for our equipment and picked up new up-armored M998 HMMWV's. After a few days we had everything we needed and we did our best protecting the unarmored M998's using sand bags. We had a brief so everyone was on the same page and we rolled out in our vehicles 650 miles to Ramadi. It was a somber moment. As we approached, the environment seemed to change. I was a saw gunner, so I had position in the back. I sat on a bench built in the back and put my saw with bi-pod out on the roof. As I covered either the three or nine switching every hour, I watched and observed. We started seeing spots where explosions obviously have gone off and the occasional vehicle that was hit and left for some reason. That fear

I mentioned in chapter one was building and I haven't even done a mission yet.

We arrived in Ramadi, Iraq Al an-bar province Camp Junction city midday. We set up our bunks and placed our gear on it. I set mine up and placed my gear on it, and I noticed a hole in the ceiling. A whole chunk out of the wall and floor, it looks like a mortar hit I said. These bunks were nice and comfy I thought, screw the hole. We didn't go right out and do missions, our leaders had to go out with them for a few days. This allowed them to get a lay of the land. We call it a left seat right seat ride with the outgoing unit. Meanwhile, we are settling in and building benches and stuff.

Places to clean our gear was important. We also bought them folding sports chairs. People even bought TV's, I just had a portable DVD player. Steel ¼ inch thick plates were welded to the soft M998's and we bolted steel plates to the sides of the M113's. The plates were 16" high with sandbags behind it,secured with 550 cord woven like a net. This turned the top opening into a fox hole that easily fit five and provided great cover. They eventually finished the left seat right seat ride and we went on our first small mission. I remember it was a relatively easy mission looking back at it. I felt the fear when my feet touched Iraq soil outside of the gate.

By September 11th, 2004 we were doing missions. Covering 1st and 2nd MEF (Marine

Expeditionary Force) is what we started off doing, and we often worked with them. We had a brief by the commander before he met with our leaders. We were about to start clearing route Michigan AKA IED alley. We started the next day rolling out the gate, taking a right in front of Camp Blue Diamond, the marine post. Taking Route Nova to Route Michigan, we started our route clearance mission, we had to wait for the outer and inner cordon to set up first.

Once set up we started looking for IEDs (Improvised Explosive Device) and clear debris. The ordnance was defused and collected in the Ace (Armored Combat Earth-mover) along with debris. Some were obvious and some camouflaged to look like bricks and curbs. We found so much and we had so many soldiers covering so no trigger man could get close to detonate, I thought. I was with a small team on the ground led by Sgt Dillon. Each day we moved further down the road and a yellow barrel caught my eye. I think that would have caught my eye the day before even from a distance, I thought. A bright yellow 50 gallon drum next to a light pole right between the fork in the road, yup that's an IED, I said.

We are even closer now and Sgt Dillon wants to position me across the street from it. I told him it was an IED. I just sucked it up and threw my camera to the guy across from me to take a picture of me, maybe my last. I made my way, keeping my distance but the position was right across the street from it. Staying far away till I got there was the least I could do. I stood there pulling security but all I could do was stare at a 50 gallon drum, and all of a sudden it exploded. Like that movie Wanted but slowed down. The barrel became this plume of black smoke. Fire began to show in the middle and I sprinted like I used to at my nana's. I saw a corner so I duck and turn, putting my back against the wall.

Aldrige was on the other side of the alley I ducked into looking, like he saw a ghost. He looks at me then this large basketball size piece of shrapnel in front of him and says, "That fucking thing flew over your head and almost took out my legs." A visual of me not ducking and losing my head flashed in my head. The only one that got hit was Skelton in the arm half a mile down the road. This IED was three 155mm artillery shells. The massive explosion left a crater in the ground that a busted water main started squirting out of the ground and the pole was barely standing. That scared the shit out of me and I decided if I go I'm not dying with fear in my heart. Fear lost its grip on me that day and I carried out my missions fearlessly.

A few days later on Aug 25th we were stuck on nova around the hospital because an enemy was firing at us. We are pulling security and I start ranting about how I could kill the enemy and was tired of waiting. Sgt Barber got tired of my cocky rant and yelled "why don't you get him then." Roger Sergeant I said. Then he says, Unruh, take his place as I open the rear troop hatch. Jumping out the back I turn and say can I take a battle buddy. He shakes his head and says Steele, go with him. I see Unruh and he looks at me like I lost my mind as we crossed paths, but maybe he felt something was going to happen looking back at it. Steele and I made our way through the orange tree grove to the end, just in time to see the MK19, an automatic grenade launcher, get a direct hit. Shit, let's head back so they can move out, I yelled.

On the way back I could hear a nine line Medical evacuation Medevac being called from a vehicle's radio. When I got back to my spot I saw a bloody knee pad in it. The knee pad was Unruh's and I asked what happened to Unruh. He was hit, they said, and I felt disoriented instantly and the sorrow made me feel sick. The sniper on the hospital had an overview of everything. The bullet entered his pelvic area bounced around and lodged into his heart. The M113 medical track quickly took him. Walker, the medic, patched Unruh up in the track as they took him to Camp Junction city. They loaded him on the Black-hawk to TQ as Walker watched. He wrote that he was a fighter and it inspired him on Unruh's memorial page. He was apparently very conscious and in good spirits from what I've read on fallenheroesmemorial.com/oif/

profiles/robertoliverunruh. We continued our route clearance mission the best we could.

At one point Lt Schultz was walking and talking on the radio and an RPG blew in front of him. Gun fire going off everywhere, I thought he was gone, then he came walking out of the smoke. Still on the radio calling shots, wow. Shoot, move, and communicate, I thought and I kept covering my sector. Being sure not to fire unless I see a weapon was our number one rule. Completing this mission was not easy. They really laid into us, but we cleared the Michigan AKA IED alley finally. Still it was very dangerous and we constantly took fire whether we stopped or blasted right through.

We got back to base and Cpt Blackmon gathered us together to tell us Unruh didn't make it, they found the round too late. I was devastated. It was my fault I should have shut up, I thought, why not me I'm just a former gang banger drug dealer. I just sat there trying to come to grips with this. Sfc Hankins sits down and says "so you're from Arizona too." I flipped out yelling and punching the wall. Fuck, fuck, fuuuck I yelled and stormed off. I didn't know he was from Arizona till then. It felt like that day somebody from AZ had to go. That day the survivor's guilt really bothered me. Being my best self is all I could do.

The next day on the 26th we decided to change things up and do simple cache searches. We did our usual brief and rolled out early morning. At around 0700 the sun is lighting everything up as we make our way. I am in the troop hatch covering our seven. I looked at the up armored M998 behind us and an explosion went off hitting it, missing us. I thought they died but as we ran up the smoke cleared, and the whole front was gone. Sgt Barber and I pulled them out and found the gunner Wenk busted up. Beach, Gutton, Madison and Weber were ok. When the daisy chained IED went off the front end popped up blowing the whole front end off. This sent the turret and gun in the gunners face but bravo had no casualties that day. The trigger man had to be in the only village on site, so we started searching and knocking on doors so our interpreter could ask questions.

I had my metal detector doing my sweep and stuck it in a septic tank under our house. Just my luck; it beeped, so I had to take a look. Tucked away close to the entrance was a mortar tube and site along with a bayonet. Thank god I didn't have to go in, I thought. I kept the site and bayonet of course, but not the tube. Our next casualty actually happened on route Michigan. It was another company called Alpha on October 11th. Russin was the M113 driver and as they blasted through at night. They slowed down, then two of the enemy's 7.62 rounds hit his head. They rained fire on us every time, but as we got better we became dangerous.

Back to Russin: he was critically hit and couldn't stop until the vehicle crashed into a building ripping the 50 cal off. The TC (Track commander) Sgt Pitts said he could hear rounds bouncing off the vehicle as they regrouped after being left behind by the convoy due to their commander Cpt Law. Lux was on the ground providing cover fire. A proper head count should always be done but was not. Rattled, they rolled out leaving him behind. It's on YouTube titled Left Behind. Lux did the best he could to stay hidden as the enemy shuffled about looking for him. He laid in the tall grass with his rifle to his chin. He was going to shoot himself if found.

They didn't find him and gave up. Now he had to find out how to get back. He had the great idea of turning on the IR (Infra Red) beacon on his Kevlar. When a marine convoy came through he stopped them by throwing his Kevlar in the road. The commander of Alpha had no intention of getting the squad or him, which is sad, but Lux got himself back to camp Junction City. Three days later on October 14, Washington, Barbret and Sgt Hawkins' vehicle got hit by an IED. All KIA (Killed In Action) except Washington, who died in January from his injuries. Now that we cleared the IED alley up, we shifted gears to cache searches which was way less fighting. We found weapons and explosive charges like artillery shells, RPG's and mines. They turned these into IEDs, which was so simple that it would be easy to reverse engineer if I wanted to, I thought. We got through November with no casualties. I'm not sure how but we did.

On December 5th Bravo Company did a boat mission with the marines while we continued our searching. A select group went with them to search this island. The island was to the North of route Nova. The boat had a mini gun and I wanted to go but wasn't picked. They took fire from the shore while searching and Ward was hit in the neck. He died shortly after. On December 8th during a dismounted patrol, Sgt Williams of Charlie company was KIA by small arms fire. Finding the explosives and weapons basically took it out of their hands. This cut the supply. We ended up finding a huge cache of 155mm artillery shells which they often hit us with like the 50 gallon drum and other crafty ideas. We started a chain of soldiers from the cache to the LMTV's (Light Medium Tactical Vehicle) and it still took three hours to load. By then it was night. Plenty of time for them to set up and I'm sure they were pissed, we took the supply. You could just feel it coming. I remember feeling very anxious and excited at the same time, the shit was about to hit the fan.

Three hundred meters before the left turn in front of Blue Diamond to get back an IED Blew. This EID blew the track off the lead M113 which takes 20 minutes in daylight to fix. This left my platoon and track on Nova. A raised stretch of road higher than the town but level with the hospital roof. Immediately an RPG fires, hitting the track ahead of us bouncing off and exploding on the floor. They rained down on us, firing from the ground, windows and roof top of the low level of the hospital. We returned fire with deadly accuracy, firing at muzzle flashes, hitting target after target. With Mk 19 shells bouncing off my Kevlar and everywhere, due to Ssg Caliboso firing like crazy, I made it through: we made it through.

Their tracers were green and ours red, and it looked like a star wars battle or fourth of July. I returned fire on each muzzle flash, being sure to stay low. Window after window of the hospital until I see flashes on the roof. I shot a five round burst and the enemy fell off the first story past the windows. I get back to firing at muzzle flashes when I see green tracers in front firing up from below. I knew right where to point my SAW (Squad Automatic Weapon) and unload five rounds, hitting the enemy. The track got fixed and we rolled out with no casualties to Junction City to do our post firefight cleaning and PMCS. We did many more cache missions finding weapons also, ultimately taking it out of their hands.

We held the first official voting in a long time, so the citizens can vote for their officials. This would liberate them, and eventually they would rebuild. They simply filed through, picked their candidate and dipped one finger in ink. That prevents them from voting again for that day. We did this with no problems from the enemy. After Ward our company Bravo didn't lose anybody, just injuries. One day I was walking to the barracks from chow with my head down, but outside the wire you can't. I see a trail of blood as I approach and it gets bigger. I get to my bay to see them cleaning up. I asked what happened and I was told Skinner blew his thumb off. Just a few days ago we gathered all this unexploded ordnance at a bunker and blew it up. The explosion was massive, it created a huge cloud. Skinner took a booster charge from an Italian mine, the initial charge that sets off the big one. The charge goes off in his hand, blowing his thumb to pieces and damaging the rest of his hand, so I was told. He went to Fort Carson Colorado to get surgery to replace his thumb with his big toe, where he waited for our arrival. Other than that craziness, Bravo was good to go. Clearing that major route allowed the marines to set up combat outposts.

On March seventh, Charlie company lost two soldiers to a VBIED (Vehicle Borne Improvised Explosive Device). Sgt Bossert and Franklin were manning a vehicle checkpoint when hit by the VBIED. We still got in fire fights here and there, but nothing like how it was in the beginning. It was like we put them in their place. Our interpreter felt it also. He tried to go see his family and was killed by the enemy. Cache missions and OP (Observation Post) missions were our racket until time to go, it seemed. Things had cooled down so I decided to get my wisdom teeth checked out because they were hurting. The dentist tells me they should have been pulled, so I say pull them then. What was I thinking? He numbed me up and began the process. With a chisel and hammer, he places the chisel and strikes. You hear a loud crack and a deep suction-like sound as your tooth is ripped out. They did all four and then stitched me up. I got Percocets and four days off. I still had to worry about the mortars that often hit our base but I got to relax.

Before you know it, August is here and we are packing up doing a left seat right seat ride. Everyone is happy because it feels like it is all over already. I don't even think we took most of our vehicles, we must have left them. A concert was planned for August 22nd 2005. What a coincidence, I thought. I turned 21 then. We had the concert and I walked around with my weapon slung, sipping away. I officially turned 21 in Ramadi, Iraq Camp Junction City. As I walked around on the outskirts of the crowd, I thought about all that had happened and all the ones we lost. The concert ended with the song "God Bless The USA" by Lee Greenwood.

Chapter 11

Resuming Life and Finding Love

We got on a plane in Kuwait, but this time it was a commercial flight. Our weapons just had to be disabled by removing the bolt and off we went, making a stop in Ireland. Ireland was a three hour layover and I just turned 21. Oh yes, I got a bottle of vodka and some Irish beer at the bar in the airport. We all did: it was time for celebration. The foam was thick like whip cream and the beer was very dark. We sat there drinking our victory beers till it was time to leave. A very long flight was ahead, so as soon as we took off I started taking shots and passed out. I woke up and we made it to Colorado Springs, CO where Ft Carson was, our final destination. The buildings for our companies and barracks for the soldiers were occupied in no time, and I had my room and my stuff from Korea. The connexes were there waiting for us so we got our stuff easily. Conte, Steele, Z and Henderson planned on going to Conte's parents house in Denver. I went with them, of course. The clubs and strip clubs were so fun to say the least. We all passed out at Conte's when the night was over.

That night I had my first combat nightmare, where I was running from the IED and the shrapnel took my head off. The second I experienced Unruh's position the day he died. In my dream I got sniped instead of him. These and other situations that we made it, though I died in my dreams. They were frequent for at least a year but meeting someone made it bearable. I never talked to anyone about it or I would have looked like Rambo at the end of first blood. STT (Sergeant's Time Training) was the very first thing we did, led by Sergeant Glick up Pikes Peak. A simple hike to get adjusted to the climate maybe, I thought to myself. The halfway point was easy and there is even a cabin to grab a bite while you relax. After the halfway point the terrain gets rougher and the air gets thinner. You can feel it. I found a path and stopped climbing. Thanks for the path because I got to a point of exhaustion. The point I was at was when I could only take ten steps at a time then rest. At the top was an oxygen bar and stuff to buy of course.

On the way back to base Sgt Glick asks me to be his wingman to keep his girl's sister occupied.

Sure, I say, so later that night he picked me up and we showed up. We got to the house and were told to let ourselves in while they were shopping. I looked around and happened to see a picture of her and her military husband! Sgt Glick, what the fuck, she's married to a Sergeant! They showed up that night I will never forget Sept 3rd. As soon as I saw Heather, time seemed to slow down for a second or two. "Did my heart love till now? Forswear it, sight! For I ne'er saw true beauty till this night" (Romeo and Juliet). She looked wide eyed at me with interest and curiosity, it seemed. It had to be obvious it was love at first sight.

She and her sister went to the room and she kept talking about me, she said. Her sister stayed in the room and we spent the rest of the night hanging out and drinking. At one point I kissed her when Sgt Glick went to pee. The kiss was short and sweet as I recall. The next day I was trying to figure out how to find this girl with no luck. I didn't go out of my way for any girl since no girl did for me. That was love. I was sure of it and I was going to find her. My roommate sees me freaking out and asks me what's going on. I tell him and he says hell, I got her number, we helped her move stuff but Glick is dating her. I got the number and it seemed meant to be but I didn't call immediately.

When I finally decided to call she knew exactly who I was. I didn't care that she was married. I was in love. We talked as well as connected on a physical and emotional level. Key word being connected. No girl ever wanted me in that way and cared to get to know me. I started having a serious relationship with her. I was there and her husband would call from Iraq, and I felt bad until I heard him make her cry. How does somebody do that? I thought. He didn't seem like a good guy.

I was ready to go watch a movie. I was still adjusting from being in a highly stressful environment, but I had to do something. I went with Heather and it felt strange as well as a bit victorious. *Get Rich or Die Trying* was in theaters and we had to see it. Watching the movie, I said holy shit out loud. I said I cashed out and went a different direction, just like this. 50 Cent didn't join the army and I didn't sell crack of course, but you get it. We both had a strong female role model though.

I started talking to Heather and spending all my free time with her and her three kids, Brandon, Makenzie and Belle. Her having three kids didn't bother me. I was in love and I was sure she loved me, too. I would stay the night and she took me to base in the morning. She even had a drawer for my things. I was sleeping over sometime in November when I got a call from Conte. Conte asks me if I can sit down and tells me that Z has died. My heart dropped to my stomach and I yelled "what the fuck happen!" Apparently another soldier, Spc Timothy

Parker that had just come back from Iraq, hit him in the head with a fire poker. He was rushed to the hospital unconscious and was pronounced dead the night of November 11. His name was Piotr Szczypka.

To make it through all the stuff we made it through and die by the hands of some coward. The guy that killed him most likely never saw combat. Now that is just fucked up. Running at Ft Carson was tough, because the air is thin due to the elevation. I got used to running there but then the winter came. Walking through snow isn't bad but slipping on black ice sucks. I would slip at the bottom of the barracks stairs every time, no matter how careful I was it seemed. We all got there towards the end of summer and it was nice till winter.

One day Heather asked me to go snow sledding with her in the mountains. We had already gone to Garden of the Gods and many other things together by then, so I was up for it. We get there and she tells me to hold on while she finds the usual spot, of course I can't wait. After five minutes I take off uphill and find a spot myself. I set the hard foam sled down and hop on. As soon as I hit the sled, I went flying downhill fast. I see a bump in the snow that was a stump. I see it and I think I'm about to go airborne. As soon as I hit a stick poked through ripping my jeans and breaking my tailbone, it could have been way worse. The board stopped and I went flying in the sitting position, feet forward across a road, as I'm flying across I looked left and saw Heather. She is just as shocked as I am. Here she is walking down the road and I come flying past her in the sitting position. I landed legs forward and knew I broke something. That was the end of that trip.

The ride back was horrible, and I felt every single bump in the road. The next day I went to sick call and they put me on profile. It took about a month to heal. She also took me to a place called the haunted tunnels of gold camp road, and that place was definitely spooky. Heather was one of few relationships I've had. Sure, I have been with many women but they didn't love me. She wanted to spend time with me and talk among everything else: that's a relationship. My ETS (End of Term in Service) was coming in March and I told her I would like her and the kids to go with me. She told me she will think about it and that her husband is coming back soon. He cheated on her a lot, I was told. Everything she told me about him was bad. We kept seeing each other and going places just because we both felt something. I guess you can call it love.

Christmas came around and I spent all my time off with Heather and the kids. I took the kids sledding down the stairs into a pile of pillows. It was crazy. We spent the rest of the time playing video games with the kids and hanging with Heather. We all went and visited my mom and picked up my dog

Phoenix. That dog was one of the many things I got when I started making money at 16. I also got some weed when we picked up Phoenix for a few days. Christmas was finally here. I had my dog, Heather, and the kids enjoying it in Colorado. The kids were in bed. It was Christmas night and they passed out from all the playing. I broke out that weed and smoked it with Heather. I usually don't usually smoke without 30 days leave, but it was Christmas break. I spent all break with them, and it was so fun playing with the kids, playing video games, and watching movies with Heather. I felt like I was right where I needed to be relationship wise and life, although she was married. This was not like me. I had never been loved. Nobody but my mom showed me love so I never gave my heart to anyone. I gave my heart to them and I didn't even know that till I looked back. Maybe they helped me heal it. Iraq severely tore me up mentally, emotionally, and spiritually.

After break we had a random pee test and I was picked. I thought oh man this it I'm done for. A dishonorable discharge takes that college money away, so no more MGI bill. I was definitely dirty. I just smoked days ago, but by some "coincidence" nothing happened. Heather's husband came back in January so all we could do is talk to each other. Boy, we talked so much. We talked and texted so much her husband found out and started looking at the phone bill. He found my number and called me, and

I answered of course. He couldn't find me. I simply told him "I don't know you man, maybe you have the wrong number." He eventually confronted her about it and she denied talking to me, thank god she didn't tell him who I was. I don't know what the article or chapter is, but I would be a dishonorable discharge if caught.

We still talked and she decided she wanted to leave with me but the kids had to finish school which didn't end till May. Her husband continued trying to find out who I am, but he was way off since I got my phone far from the post. She wanted to get away from him so we left together on March 10th 2006. We planned on getting set up and getting them after school was over. My dog Phoenix died a month before I got out of the army, so the kids were told when they asked where she was. We set up at my moms house in the largest room we called the family room, and headed on back to get the kids at the end of May.

We got a hotel room in Colorado because she said I should wait there. She didn't want anyone fighting. There I am waiting, after about four hours I figure I should call to see if she is ok. I get hung up on over and over until her phone is turned off. So many things started racing through my mind, like did he hurt her, is she just packing or what if she is not coming back. I was pacing back and forth heartbroken. A day ago it felt so full and now my heart felt empty. I didn't sleep at all that night. The next day I didn't know what to do so I ordered a pizza after trying to call one more time. How could this be happening? This is so bad, like a scene from a sad movie. I came to the conclusion it was over and started making plans to get myself home from Colorado, but just then she called. I was so happy I didn't bother asking what the heck happened. She told me she was on the way and to be ready to go. To this day I still don't know or understand what happened.

We made our way to Phoenix AZ, Heather, me and the kids. My mom raised four of us and she was on her own most of the time. Heather and I could raise three together easily. We got to my mom's house and moved into that large room. We had the full size bed and the kids had the full size air mattress. The entertainment center was set up for the kids and we had a computer as well in front of our bed. We had all summer to settle in and enroll the kids, Brandon and Mackenzie, into Starlight Park Elementary School, the same school I went to. Belle was still a baby, she wouldn't start for a while. Heather, Gabe and I went to The Grand Canyon and stayed a few days. It was a blast. I am resourceful, much like my mom, so raising a family financially was easy. I had plenty of money but didn't rely on it alone. I still had her sign up for food stamps, and we loved getting Tuna Helper, Hamburger Helper every helper they had. We got the kids their uniforms for school since they required them to wear white shirts and blue shorts.

Sometime in September they started school. They were in third and fourth grade by then. Belle was starting to talk, still long from going since she wasn't even one. Heather got a job as a dental assistant and I got a job as an appliance installer for a company named Barlow. Ray Zepeda AKA DJ Ray, my mom's friend, got me the job and we became a two man team. I learned so much on that job, and we installed everything from microwaves to the huge sub zero fridges. I was the one unpacking and moving everything we installed, so imagine 5'8" 165lb me moving a 900 lb fridge with a dolly. He was the brain and I the brawn for the job, even though he was 6' 3". I still get a few bucks to be the brawn and move some DJ equipment here and there.

After a while Heather got tired of us living at my mom's. We did need more space. We had barbecues while the kids played in the plastic pool in the backyard which was fun, but space was needed. She got to the point where she was belligerent toward me about the subject, I figured she was in the right. I went ahead and moved us into an apartment, and it was easy since we didn't have much. We immediately got furniture for the bedrooms, living room and kitchen we fully stocked. The furniture was very nice, we picked it out regardless of price. I stopped working for Barlow and settled for something easier with more room for promotion, so I worked at Target. I started college for a Bachelor's degree in business management at University of Phoenix, hoping maybe one day I'd be a manager at Target.

My grades were pretty good and I got the hang of powerpoint presentations, writing papers and leading the team often. I had a beautiful woman that loved me, the kids loved and looked up to me. I felt the love and felt on top of the world. We went to the pool at the apartments often, and when we got back I would cook up some food or Heather made something. We also went salt river rafting and took trips often. I worked part-time and Heather worked full-time as a dental assistant. I spent a lot of time home alone, which is not good for a soldier with fresh PTSD (Post Traumatic Stress Disorder). When we had an argument and she went to work I would not stop calling to say sorry. I did this because I wasn't right in the head. She was let go because of me. I feel bad about it looking back at the situation. She didn't need to have a job for money. I added her to my account and she was set. I had her back 100%.

I was working hard at Target and going to college all was well, but I started watching porn and Heather didn't like that. I guess I had a high sex drive and her a low one because it seemed I always initiated the activity, but it was always amazing for us. She had her tubes tied already so kids were not going to happen but I was open to it. I had a bit of a problem. I figured maybe she isn't as attracted to me as I am to her. It got to a point were I couldn't have enough of her, and we must have had sex 365 times or more in one year. I suspect she may have been bipolar because we had it so good but she would flip out from time to time. I figured it was my fault, so whatever it was, I tried my best to fix it. I didn't hang out with my old friends, I just spent all my time with her providing whatever they needed. I was starting to feel this distance between us and it started with us not connecting like we used to. This led to me waiting for her to tell me if anything is wrong or at least talk to me. I started to feel unwanted or unattractive.

We got drunk one night and she flipped and left, taking her kids to this couple's apartment to play with their kids while they drank. I had to go to work tomorrow but I couldn't go. She didn't come back that night so my brother would give me a ride to work. The next morning, Gabe picks me up and on the way to his truck I hear scuffling on the patio. Heather is hiding in the shadows. Wow, I had the only key, she left hers inside. I let her in and she told me this crazy story about what happened, but I have no idea what really happened that night.

Our relationship was getting bad. She would start hitting me repeatedly and one time I struck her with my backhand on accident in defense once. I got arrested for it when she called the cops and she dropped the charges. Nobody should be hitting their loved one. I blamed it all on not going well on myself, the porn, or maybe I'm just not good enough for her. I felt like enough before and I felt love in the beginning, so what can I do to fix this. I decided to tell her we would move out to a house and have even more space and things. I just wanted desperately for things to work with the only woman to ever show me real love, a rare thing for me. She was very surprised and happy and soon we moved. The kids each had their own room for once and I got them all game systems. The girls shared one. We had three TV's and systems hooked up playing Battlefield together, including Heather, Brandon and I at one point.

I decided to visit Unruh's grave so I googled his full name and found out his mom died soon after seeing his dead body, but the article said where he was buried. The article said that she died of a broken heart which was strange to me. He was in The Broken Heart Battalion after all. I asked Heather to take me and she was mad about the three hour drive. I wasn't ready to explain it to her. The first time I went I felt this overwhelming feeling of guilt. I just put my hand on the nameplate and trembled and cried in silence as Heather sneered. I think about him when I listen to "See You Again" by Wiz Khalifa. Bright side is we had BBQ on the grill out back, beer in the fridge,and plenty of money in the bank. It seemed I fixed the situation. Besides, the dream of me taking his spot that day started popping up a little more often. I was good.

I went to work early in the morning in her car, was back before 12, and she spent the day at home so she had time to do what she wanted. When I would get home she would find something to argue about almost as if she was trying to find something to hate. It was a nightmare. Things were great then boom, not good, and it perplexed the hell out of me. I was a wreck to where my stomach was giving me problems. It was nerves I guess. Dealing and worrying about her, my school and work, and trying to make progress all at once must have had me at the brink of a nervous breakdown. We started arguing again and most of the time I would spend half the night apologizing no matter what I thought.

One day Heather tells me her grandmother is dying and she needs to take the kids to see her in her home town Rock Hill, SC. I thought nothing of it. I had to work and I couldn't go. I told her and the kids I love them and off they went in her car. I was thinking maybe this will give us space so we could appreciate each other. After a week of walking to work and sometimes getting a ride, also no answer from her when I call, I begin to worry. I started to see pics on Facebook of her with her husband having fun, then I got a call from her. Honestly, I don't remember what I said to her but I forgave her and he was done with leave back on base. I have no idea what happened between them during that time and

I never will. We don't talk much. Now we talk from time to time when she needs help. The next place they were going with her mom was Myrtle Beach, SC so I got myself a ticket and flew down there to see her.

Chapter 12

Proposing and Trying to Fix This

I should have been upset, but I truly thought this was all my fault. I wasn't good enough to her, I thought. I got to Myrtle Beach and took a taxi to Compass Cove with nothing but the clothes on my back. I can buy what I need later. I get there and start talking to her, and for some reason I remember her being mad at me. She looked so sexy in her swimsuit is all I can remember thinking as she stomped around the hotel room ranting. I was there so I took everyone out to eat after Heather went on a walk with me to the beach at night.

While walking and talking on the beach I got to the subject of marriage and having kids. She was definitely open to the possibility, I realized, so I dropped to one knee and proposed with a ring from the airport. I was on my knee in the moonlight and she said yes! We were so happy and I remember feeling so amazing. We couldn't legally marry until she divorced, but we would take care of it. I was going to get her a better ring but it did the job for now. After Myrtle Beach I was hoping everything would be all good. "I'm a sucker for love" as 2Pac once said. We got back to Phoenix and I got her a way better ring. I put it on the ice in her drink at Margaritaville and it fell to the bottom. I said what's that and we fished it out with the drink stirrer. She loved it. I told her the wedding ring would be better than that. We found a fertility clinic that recommended in vitro fertilization after telling us her tubes could not be reattached. This was going to cost 30,000 dollars, and after talking to Heather, we agreed to do it.

They asked me how I planned to finance. I never finance, I told them, and paid it with a check. The process involved her taking a drug to produce many eggs, then they extract them. They put my sperm and the eggs together to make many embryos, which you pay to keep frozen. For the final stage she was put on a drug to prepare her for the progesterone procedure. I think this was very hard on her. The first time it didn't take, but five grand more gave us a second try which took. This was confirmed by a pregnancy test we bought. She just needed to have an ultrasound done every three months. Meanwhile we were talking to an army recruiter and I ended up enlisting as a 12B combat engineer once again.

I got three years and chose my duty station, Ft. Benning GA and I got Gabriel to join. Funny thing is after three years I missed being a soldier so I came back. Obama was the main reason. He said we were pulling out of Iraq, but we didn't, and all of a sudden my son is struggling with common core math that skips steps. He did nothing good for the country.

We happen to see this lady selling red nose pit bulls on the side of the road and that's where I got Simba. Simba had big paws so I knew he would get big. We continued our life at the house which we planned on renting out when we left. I barbecued while the kids played in the pool, and Heather started packing stuff up since she found a good house for us to rent. This place was in Phenix City, Alabama which almost spelled like Phoenix and I thought that was pretty cool. I flew her dad in to drive the moving truck since my report date was in January. We had our Christmas celebration and left soon after leaving my mom in charge of finding a tenant. We got there and the yellow house was on a hill at the end of a driveway in a cul-de-sac with an above ground pool. It was so cool. We could not get the truck up there so we took everything up by dolly or hand. Heather took things up in her Ford Fusion also, so it was done quickly even though we had a lot by then. We set up all our furniture and unpacked as much as we could because the next day I had to report to my unit with my orders.

The 11th Engineer Battalion (Eng Bn) 36th Engineer Brigade on Fort Benning Georgia was my new unit January 2009. I reported to Bn and Sgt. Diaz took me to the unit they wanted me in. Very different patch than I was used to, and the uniforms were now a digital pattern of gray with velcro they called ACU (Army Combat Uniform). The patch I was used to was 2nd ID which was an Indian head. This was a freaking seahorse, I thought. Oh no, I went to a pogue unit again like Korea. I got to HHC (Higher Headquarters Company) and I was greeted by Sgt Walters. She had this serious vibe to her and is pretty, I remember thinking.

I met the commander which went pretty smoothly. All there was left was the 1st Sgt. That did not go smoothly at all. He was glaring at me like he was mad. He was ready to blow, I thought. All I remember is him yelling at me telling me to get out, and as soon as I come out Sgt Walters is in my face yelling at me telling me to get in the front leaning rest and give her thirty push ups. So I started then and she started yelling at me because I was smiling. I was thinking hell yes, I'm back. I was assigned a room at the barracks which worked for me since my house was 30 minutes away in Phenix City, Alabama. In the army paperwork I was single. That meant no extra money for housing, but I didn't need it: I had plenty of money.

It didn't take long for us to get the house situated and meet the one neighbor up there. He was in the special forces. He was no longer in the elite group due to three purple hearts, and apparently that makes you non deployable, but he was an instructor for airborne. Our wives hung out often and we went over and partied with them. Heather was happy and everything was perfect. We were talking and connecting like we used to in the beginning. It felt like a dream. We did our ultrasound and found that she had what was called a blighted ovum, it was developing then stopped. It was just a black blob. We were so sad about this and cried together for quite a while.

As for the unit, I truly thought I fell into a dead end pogue unit and was not going to get to do anything cool since they put me in HHC 2nd platoon. All I did was run errands and help supply them with their shipments that came in. It was Specialist Moreno doing boring nine to five work. I did the occasional CQ duty which was 24 on then 24 hrs off, but other than that, for six months that was all I did. July comes around. Of course I got ribs on the grill and the huge above ground pool is clean. The kids are jumping off the big 12 foot trampoline into the pool and also screaming "Jimmy Fly Snooka from the top rope!" while smacking my elbow. I had many people over, at least ten or more, and we just grilled and drank. I felt like king of the hill because you can literally see through the trees to the rest of the street down hill. That was most weekends for us. Sat outside and just talked, just Heather and I, about anything. I truly loved it. I felt like I was being a good man for her and her kids. She was finally divorced but we didn't marry yet. She took me to work in her Ford Fusion and picked me up until I got me a truck.I drove a navy blue 95 Dodge Dakota to home and back. I usually didn't get a car, but I had to get to work and she needed to take the kids to school.

Chapter 13

Being a Good Man for My Family

Heather and I were good to go, and the kids were well taken care of, dare I say spoiled. It is not good to spoil your children, but I was still a kid myself, just 26 or so. Heather's white and her family was ok with her seeing me, a Mexican guy. At least that's the impression I got, but I felt a tension. I never really worried about being a good man. I just did whatever for me and never provided or set an example for somebody. Now I was providing for my family. Heather was loving on me again, and the kids looked up to me with love also. I loved being a good man for them and providing an excellent lifestyle for us. Back at the company, the command changed completely. The commander was Cpt. Stephanie Radford, and the 1st Sergeant Charles Fuller. Our Executive Officer (XO) was into boxing and tasked Ssg Richard Fuentes to put a boxing team together to fight at the 2009 Ft. Hood boxing competition, being held from September 8th to the 15th. I volunteered of course, and others from my company did too once word got out at Bn formation.

Those that volunteered reported to a small building that we turned into a boxing gym for PT. With eight punching bags, two speed bags, mirrors on the walls, and equipment, we turned it into a hell of a training area. It was July 5th or 6th and it was in September, so we had two months to train and I trained like I was preparing to fight a demon. The volunteers were Freddie Grant, Ruby Ruiz, me, and a few others. We had equipment due to our supply (S-4), which they ordered for us. For two months we trained until it was time to go. Rounds on the bag, shadow boxing in the mirror, and jump roping with some running is basically all we did to prepare for the 2009 Ft. Hood Boxing Competition.

I was training, doing my boring pogue job and Heather had the kids home by the time I was done. Dinner was amazing, especially her meatloaf. No arguing, just love and happiness, like Al Green once said. We went bowling at the bowling alley down the road as a family, and I often took her out to nice restaurants. We called it date night. I would buy her nice things to wear for our dates as well: earrings, dresses and shoes too. We had a very good life. So good, in fact, that Heather decided to ditch the

Fusion and get a 2010 v6 Ford Mustang with dual exhaust. It was a very nice car.

September was here and we had our flights to Texas booked. Even our rooms, courtesy of the Army. We got there and rented a car with the card Ssg Fuentes was given, since the rooms were taken care of. We went to Ft. Hood to register. Once there we filled out our papers and weighed in so they could pair us against fighters in our weight class. We had got there a few days early so we had time to check out the area around Ft. Hood, Texas. We went to a restaurant, and after the delicious meal, someone said we should check out the strip club. Next thing you know we are all at the strip club, and I remember thinking Heather would somehow find out. The fight was a few days away. One night of fun wasn't going to make a difference.

The next day we were good to go since nobody really drank, so we got back on schedule. A short run in the morning and stretching the rest of the day is what I did until the fight. I weighed in at 165. I can't remember the name of the fighter they paired me with but he was a black guy about my size. All of our team got paired with an opponent in their weight class. During this trip to the Fort Hood Boxing Competition, Ruby Ruiz and Freddy Grant started to get to know each other on a personal level. This led to them getting married soon after we came back. The day of the fight was here. I was the last to fight so I recorded everyone's match. Grant was first to fight and you could tell he fought before by his balance. He was swinging away, trying to take this guy's head off. Grant was black and his style seemed like Tyson, so I said damn Tyson, giving him a high five after. Ruby was next. She was Mexican so I told her, kick her butt Roberta Duran, a female version of a Mexican boxer Roberto Duran. She came out with punches in bunches. She used to box and it was obvious. Each round she came out and just pressured her opponent. The other fighter just couldn't keep up and just kept getting hit.

It was only three rounds per fight and one fight a day, but that fighter was gonna feel it tomorrow, I remember thinking. We had other fighters with us but I can't remember their names. All I remember thinking is wow, we should have spared a bit. The balance and coordination was very bad for the other fighters, but oh well. I recorded so they can see and learn what they did wrong.

My fight was coming up next so I went ahead and stopped recording. I was in the back getting my hands wrapped and checked by the referee and shadow boxing to warm up my arms. My fight was not immediate. I still had a few fights to go until my match, so there was no rush. Since I recorded everyone's, mine was being recorded. I came out when the fight before mine was beginning and watched them fight while I waited my turn. The fight was two featherweight fighters going at it, throwing many blows. We stepped into the ring and my team was recording this fight which is on YouTube. If you google #sapper_ninja you can find it.

They announced us and we began. Immediately I could tell I was way faster and more balanced than him as I dodged punches and came right back in, landing left and right hooks, overwhelming him. I wasn't putting on pressure or throwing a bunch of blows, I was consistently landing blow after blow. Then he collapsed to the floor and I was sent to the corner while the referee did a ten count. After the count we got back to fighting he would swing, and I just slightly moved and I hit him. I kept doing this, landing a massive left at one point and his knees buckled. He was in trouble and he knew it in his eyes. He started swinging away and I kept doing my thing until I landed another left as he missed with his right. His fist went past my face over my shoulder and he pulled down just as I connected with my fist, dislocating my left shoulder because he pulled down. I lost that fight but he should not have pulled down like that.

Chapter 14

Joining Recon PLT

I got back to the company at Fort Benning just in time, it seemed. They wanted soldiers for a new platoon. They wanted only the best soldiers in this platoon called Recon Platoon, and they announced they would choose later that day. My shoulder was in a sling but I wanted that platoon badly, so I went right to the First Sergeant to ask if I could get on. He told me to report to the platoon sergeant SSG Herman, a Sapper Ranger, meaning he completed Sapper and Ranger school. I was told he was in the motor pool with our new M998s, so I went to report to him. He had a layout and inventory going on so I started helping. As I was helping he started telling us that he will start sending us to as many schools as possible. I immediately asked for airborne and he said done, just show me you deserve it. We finished inventory and he signed for our new equipment. PT in Recon was definitely harder, but that and other things set us apart.

After PT I was told I had to report to Airborne School at the beginning of October. Hell yeah, it looked like my second time around was working out perfectly. He gave me my packing list and I had a week till reporting. I was set and ready to go. I was ready to advance as far as possible this time around. We decided to play soccer for Sergeant's Time Training on Thursday. It was a good game. As we were walking off the field I rolled my ankle badly, stepping on a rock wrong or something. I just remember it was way swollen and I had to report in two days. I went and reported as scheduled, swollen ankle and all.I just had to hide it or get recycled to the next class.

The PT test was first. All I had to do was pass the run, I thought to myself. We did push ups and sit ups, just having somebody hold my ankle hurt. The run was the final event; those who didn't pass their standard were turned away. I wore pants to hide my ankle and off I went running, using mind over matter: if you don't mind it doesn't matter. Looking at me you could not tell, but I was in excruciating pain. There was no freaking way this opportunity was getting away from me.

When I finished, I went to my room at the barracks. I elevated my leg on the wall groaning in pain. We didn't have anything else till the next day, so we spent the day recovering. I had ibuprofen to help with the swelling, along with elevating and wrapping my ankle with one of my hand wraps from boxing. First week was ground week, which is learning your four points of contact and how to fall properly. An instructor we called "a black hat," because they all wore black hats, demonstrated for us. He jumped off a very tall podium and fell perfectly, allowing him to absorb the fall by popping up somehow; this was known as a PLF Parachute Landing Fall.

We did a PT run every morning at a fast pace and those that fell out got recycled. For a week we practiced on a three foot platform. We also did mock jumps from a separate platform that looked like the plane a bit. We basically went through the whole process of jumping and landing from the ground up, literally. Each morning run my ankle was feeling a bit better. I just had to keep taking care of it every chance I got. I got through ground week by hiding my ankle, elevating it against the wall, and taking ibuprofen. Next was Tower week, which was no different except all these mock jumps were from the 34 foot and 250 foot towers. Without a doubt we recycled many that week.

The runs got faster, and we lost a little more as a used mind over matter to just get it done. I'm headstrong like mom apparently. The mock jumps from the towers were going through the commands then jumping on a zip-line to simulate how the plane moves forward as we jump leaving us. I liked tower week. I was taking a ride at an amusement park all day for a week. I just had to push through the run. The 34 foot tower was OK but I loved the 250 foot tower. It was taller but the ride was longer. At last the final week called jump week was here and I was so relieved. There was less running in jump week. Now I can heal faster.

In jump week we got up, went to chow and ran to the airfield at about 0500. This week was not as fun as Tower week. I definitely waited a long time to ride the ride once a day. In tower week it was 20 or more a day. Sometimes I was standing and sometimes I would be on the bench, but standing bothered my ankle because it was not fully healed. A black hat noticed that I started favoring my injured ankle and asked me if I was injured. I told him I just had a small rock in my boot, and I will get it later. They will recycle you if you're injured and I was so close. My first jump, finally. I was very first, watching the ground below at about 1300 feet in a C130 for 20 minutes, ready to jump.

The green light flashed and he said go. I took one step and jumped right out, yelling airborne. I immediately check the cords and they are a little tangled, but I open it quickly. Free fall was like three seconds at the most since this was a static line jump, the first of five to graduate. Jumping out of a plane was fun. Anything that challenged me or made my heart race made me feel alive. I graduated Airborne school with 100% grit and maximum effort at risk of sounding like Deadpool. I returned to the company to find that Ssg Herman decided to put me in the PLDC (Primary Leader Development Course) to be sergeant. This was an easy break after what my ankle went through. It was like going to a classroom and learning the ins and outs of leadership, which is mostly soldier accountability, counseling, and forms along with many regulations.

This class environment was easy. I just had to study and pay attention in class as well as live there. We ended PLDC with an FTX (Field Training Exercise). This was us going out to the field and setting up tactically. After setting up and assessing the area we took turns being in charge of groups for some war games. That school was like a big break. I think it was the calm before the storm. I had another one coming. Air Assault School is a very tough school physically and mentally without a doubt for two simple reasons. You had to run everywhere you went saying Air Assault, and the sling loads had many details that had to be learned and later inspected to pass. We did a lot of repelling from the tower and helicopter. It was so much info but I had a great memory. One day they decided to quiz someone and it was me. Well, I volunteered and got every question right. That shocked them. It was an interesting school. On the last day we did sling load inspections and I barely passed.

Chapter 15

Advancing Further in the Army

I finally got to go home with my fiance and kids. Heather and I celebrate by having a family movie night and some drinks. She is looking as beautiful as ever. Heather is 5'3", brunette, with piercing blue eyes, and is very sexy. We had all our friends over the following weekend since I was gone for a while. It was the usual we provided most of the beer. I was back with my family, spending time with them, going out to eat, and going to the family events on base. Heather and I had a great relationship that hit a rough patch and I felt we passed that. Even as I write I'm not sure I can put in words how happy I was, maybe bliss. We continued our date nights, finding different places to go eat. Everywhere from White Castle to an expensive five-star restaurant we just went. One time she got food poisoning and spent Valentine's night puking.

I passed the promotion board soon after getting back. Within no time I was Sgt Moreno and I was put in charge of a squad. It was a bit of a transition, going from battle buddy to boss, but I handled it. Leading was easy for me since I often lead a team while attending University of Phoenix; not tactically, but using soldiers to complete a mission is the same thing. All this happened within 2009. Now that was progress to advance that fast. In our spare time, Recon trained for schools, so next I wanted Sapper School. I started learning more advanced Combat Engineer skills. The platoon Sgt saw me looking into training manuals. The info is there, I just had to find it. Later that day he asked who wants to go to the 2010 Best Sapper Competition at Ft Leonard Wood, Missouri in April. Nobody wanted to but me, so I started training with him and figured out how to do my walk on papers. There are slots given, but If you're willing to take a chance, some soldiers don't report leaving a slot open.

We trained on many things Engineer and Infantry related. We didn't know what to expect so we covered everything we could. December came and went. Heather and our kids had a blast on Christmas and I got her blue zircon jewelry to match her eyes for her birthday. Warmer weather was on its way, so we trained a bit more. The competition was two months away, and we trained so much by then that we started preparing for the road march we knew was going to be part of the competition. We did this with 35 pound ruck sacks for miles and miles, just walking. The plan was to go compete and then try to be a walk-on for the next class which I was told was totally possible, the Platoon sergeant said.

We arrived at Fort Leonard Wood Missouri on April 20, 2010 and reported for the competition already registered. The competition was going to be 72 hours straight with little sleep and various events to test our grit. The first event was a quiz to test how much we knew about our field of expertise. This test had many demo calculations and other various Combat Engineer questions. How well your team did on this test gave your team a number, and our team was three, so we did well. The next day the competition started with a PT test and boot run in full uniform wearing IBA (Individual Body Armor) and carrying a dummy rifle for three miles. How your team did in each event determined how many points you got. Next we went to TA (Training Area) 250, which is just a big lake. On the shore of TA 250 we had to poncho raft our rucks together and have it inspected for points. I remember fog over the water of the lake. It looked so eerie.

The Chinook arrived and each team got on with their poncho rafts. We went to the very center of the lake and were told which way to go when we helocast. We jumped out ten feet above the lake and started toward the finish line. All this is point oriented. Once there, we open our poncho raft and take our gear to the next station. These stations were live demo stations and land navigation stations. The live demo stations were awesome. As for the land navigation, it was basically a walk in the park, other than the fact our rucks had to weigh a minimum of 70 pounds. The rucks were heavy and I had the radio that wasn't included when weighed in. Until that day, I never did an actual charge on an I beam, and all the other charges we got tested on. I felt like I was dreaming. I was doing some real bad ass shit and the only limit was myself. I was hungry for progress.

It was getting dark and all stations were done so we returned to the huge tent we all stayed in. Once everyone was done we started night land navigation. This was my favorite event looking back at it. We had a card with several locations that had to be hole punched at each one. With a compass and map we plotted our course, using pace count to keep track of distance. The first point we found was so amazing. We just followed the basics and boom here we are. I took a rest because I was carrying like 80 pounds. Herman wasn't having it. He said get up, let's go, and we kept finding points. At one point I was going uphill with this monster of a ruck. The sun was up. We turned in our card long ago and got rest while others searched so we were ready for day two.

While we were waiting for others to finish, I got a tap on my shoulder and a "hello Sgt Moreno." I turn around to see LTC Mize. He was the commandant I think, but I said "hey sir, you remember me?" He said "of course Sgt Moreno, how are you doing?" Day two was more demo and obstacle courses I think. I remember making my first platter charge and hitting the target far away. I got a huge smile on my face as I write this. All these courses and stations and we still had the 12 mile ruck march at night. They only took top ten to run the final obstacle course and we didn't make it. I went to the sergeant in charge and asked to see where we were placed. It was 12 out of 33 teams. Our Lieutenant (Lt) Beeman and his teammate Captain Gibbs made the top ten, so we watched. The grueling competition was over and we went back to Fort Benning, Georgia.

I was back home with my beautiful wife. I gave her a big hug and kiss and told her I was going to Sapper School. Less than a week later I'm back with everything on the packing list and working on paperwork. The sergeant that was in charge of the competition was there and recognized me, so that was a plus. There were limited open slots, they told us, and had to narrow down who went. To do this, they first asked who was in Best Sapper, and I stepped forward with a big smile on my face and said me. I was in for the win. Yes, 28 days of hell was coming my way. I also recognized one of the instructors was in Iraq when I was there in Alpha company I think. The course consisted of two phases. Phase one was advanced medical, land navigation, demolitions, air and mountaineering, landmines and foreign weapons. My favorite weapon is the West German G3. It looked nice. There were very advanced things they taught us in this phase, and it is all in our Sapper Handbook if we forgot.

We did a 12 mile ruck march with 35 pounds timed, and we moved on to phase two which we started with boat run PT. Everybody got an unbroken chem-light necklace and was told if it broke, you're slacking. We surrounded the zodiac and grabbed it on the command lay ho heave. We picked it up and off we ran. There was an instructor in the boat and one on the floor. Long story short, it was a brutal PT session. Running with it on our shoulder, benching and curling it until muscle failure. Lt Beeman was there too I forgot to mention. My chem light got snapped and I was devastated, but it made me push harder to prove him wrong. In this phase we learned patrolling and we got to practice the advanced demo we learned hands on. Basically it felt like we were making homemade explosives at one point.

The kill class was my favorite cause it was so funny. The instructor showed us how to kill a rabbit and a chicken, as well as how to cook them. The funny part was when the rabbit was killed, the sight of blood dropped three people and I couldn't stop laughing. After the kill class we moved on to the last week's Field Exercise, which was go or no go. The reason being in real combat, if you mess up, people die. Rule number one was don't get lost, rule number two was to look cool, and rule number three was if you get lost... look cool is what we were told before the exercise began. Makes sense to me that the leader must keep their cool at all times and should always know their location.

After doing so well I ended up failing patrolling but did so well they gave me the opportunity to stay for the next class in a week. I took it and called my Heather to talk to her and let her know what was going on. I joke about it and say Sapper School was so nice I did it twice. Heather picked me up from Sapper School with the kids and we ate at GW's pizza, getting the huge pizza they were famous for. I made it back home with my beautiful woman and went to Myrtle Beach, South Carolina with her kids and Gabriel for the fourth of July. I got my brother to join when I joined the army the second time, and he was on leave. My 27th birthday was amazing as usual, which was back at the beach again. We went there so much. We went camping as a family at Kings Pond for a few days. Cooking dinner and breakfast on the hot plate for the family was nice. I took Brandon to the MOUT site I found on my map and we had air-soft wars.

The sergeant major changed one day in September and it was none other than Sergeant

Major John W Etter. He was the first sergeant of Charlie Company in Iraq. He walked up to Recon in formation one day and said "who wants to go to Ranger School?" I immediately raised my hand and made two of my soldiers raise their hand, and just like that we got slots for Ranger School. We got slots which guaranteed our attendance, so we just had to have everything needed and train for the class in January. Oktoberfest was going on at Fort Benning, so we went and had so much fun, and even had a funny video made of us. The food at the fest was so good and free for all. Free isn't always the best, but this time it was.

I really loved being a good dad to my kids and being a good man to my soon to be wife. Every Halloween, Heather and I got in Costumes along with the kids and went trick or treating. This time I was a ninja in multi camouflage and she was something sexy as usual. She was just scorching hot sexy. I was a very lucky man. Thanksgiving was an epic feast as usual. We both would cook the feast for the family, and we both were good at cooking. Christmas was here once again, and Heather's birthday also. She was one year older than me so I jokingly called her my old lady. On December 28th, 2010, one week away from reporting to Ranger School, I was driving on Fort Benning. The truck in front of me seemed to hop off the road, so I called my wife on speakerphone to tell her. Before she could tell me to pull over, I blacked out and started having a seizure. A fellow soldier saw me veer off road toward a tree and she pulled me out and stopped my truck thank god. I woke up in the hospital after my MRI.

Chapter 16

The AVM Ends My Commercial Break of Happiness

In Deadpool, he says "life is an endless series of train wrecks with only brief commercial-like breaks of happiness." This was the end of my commercial break of happiness. I was told that I had an AVM (ArterioVenous Malformation) the size of a large plum in the right temporal lobe of my brain. An AVM is a tangled cluster of blood vessels that can appear anywhere in your body. Mine was one in a billion in my brain, what the heck. Within weeks Heather and I got married, because if I died, I wanted her to be ok. I was willing to die for her I guess you could say. I was at Emory University Hospital in Atlanta, Georgia talking to Dr. Daniel Barrow. He had many awards. I knew I was in good hands. He scheduled my angiogram to map out the AVM and it definitely wasn't pleasant. They inserted a tube up my femoral artery in my groin area, and the destination was my brain. I could feel it and started freaking out. They gave me morphine and I was giggling, watching the ink shoot out and map the AVM on the screen.

When I got home I prayed and started a relationship with God, asking to get me through this for my family. After mapping it and coming up with a plan, we met again. He explained the severity of this thing and told me if it ruptures I have a 70 to 80 percent chance of dying. He gave me the option of living with it or getting it out. I didn't hesitate, I said take it out. We scheduled my brain surgery for March 10th, 2011. They prepped me for surgery and I just had to wait a bit in the hospital bed. I felt pressure building in the right side of my brain, causing pain. I told the nurse something was wrong, but she took it lightly. I told her again and she asked my pain level one to ten. I say ten. She gives me morphine and asks again. I say seven but something is wrong. In a last ditch effort I grab the nurse with both hands, pull her to me, and say "fucking save me I'm dying!!" I black out because I had a stroke.

Heather was brought papers for an emergency craniotomy and she was told I ruptured, flooding my second, third and fourth ventricles. She signed the papers and they rushed me into surgery. I woke to see Dr. Barrow confidently crossing his arms, smiling at the foot of my bed. Surrounding my bed was my family, Heather's side, and my side, but I did not recognize any of them but my mom and Heather. I realized I could not feel or move my legs but I didn't acknowledge it. I just stayed positive. I also had blood in my right eye. The pressure ruptured a blood vessel as well as half my peripheral vision gone to the left in both eyes.

Like the bride in Kill Bill, they started working. I guess my brain had made the connection to "move your big toe." Thirty percent of my brain had to be removed along with the AVM, so my cognitive abilities were very bad. I also had long and short-term memory loss. As soon as my legs started working I was up working on my mobility, which was also very bad from balance to coordination. I must have looked like a zombie, barely walking with staples on the side of my head weighing about 125 lbs at one point, not my normal 165 lbs. At one point they got tired of it and chained me to the bed, but by then I was getting stronger and I broke the rails on the bed. I was pulling everything out of me: IV, feeding tube, and catheter even, almost the drain tube in my head, so they put a vest on me. This vest had things on that kept me occupied like a child. Looking back, it was strange. I developed an infection in my brain so they put a PICC line in my brachial artery under my arm. This allowed them to deliver the antibiotics better and kill the infection since it was in my brain.

I had all this stuff in me and on top of that a clot formed in the front left side of my brain.

I was put out while they sliced right above my hairline and went in, and I woke to them beginning to sew up my head. I was strong through this until this point. I just started crying uncontrollably, thinking why is this happening to me. Three months ago I was on top of the world and I knew what direction my life was headed, now I had no idea. I just knew I had a huge cross to carry. My brain was getting better because my speech was getting better. I didn't think to document any of this, but I should have. While in the hospital I started having combat nightmares again, like I was when I first got back from Iraq. I had to get out of there. They were frequent for a year or two and now came in only dark times for some reason. I was 27 and God did not let the Grim Reaper take me. I feel like I'm in God's 27 club. Robert Johnson is the first member of something called the 27 club, but it has to do with dying at 27 like him and other rock stars.

I finally got out of the hospital after 35 to 40 days, and I knew how messed up I was when helping Heather gardening. I couldn't even dig a hole and put the plant in without getting very tired. I had a paper from regional rehab where I wrote about what I did each hour to remember what I did that day. They helped me recover. I had surgery done on my right eye to remove the blood from the ruptured vessel.

The crazy thing is I woke up during it. I woke and listened to them talk for a bit and went back under. I went back to work a week after getting out and getting surgery when they saw me and they looked horrified. I was 130lbs, pale, with a red bloodshot eye and staples in the right side of my head. I did PT with them the best I could. If I was running, I would go until I couldn't and walk back.

I consistently made progress until one day I was in great shape, once again leading a platoon run, and what do you know, I had a seizure. Until then I was able to hide the fact I was way messed up still. I would task out my soldiers after PT and stay in my room study once I realized I forgot so much military knowledge. I wasn't sure exactly what I forgot, so I frantically went over everything I could. My cognitive ability was still bad so I started attending University of Phoenix again under the Americans with Disabilities Act and picked up where I left off. It was definitely a long hard road ahead of me but "a journey of a thousand miles begins with a single step" (Confucius). The army started my medical board process but I planned getting as much done as possible before they finished. In my first class, I was so bad at writing papers and powerpoint presentations, but as time went on I got better, and so did my cognitive abilities. My 28th birthday came around and I couldn't be happier.

Only thing was Heather seemed to treat me like she didn't like me. I came to the conclusion that

I must not have been a good guy because I could not remember much. I remember talking with our friends and they called me a great guy. I said by the way she treats me, I don't think so. Her attitude was like the dentist's wife in *The Hangover,* domineering and uncaring, but a great movie by the way. In school I was doing very well, and at one point this girl Cham Rodgers said damn, I wish I could do a presentation like that. I was good at it. I simply had one rule: know your material. The way I organized my powerpoint and cue cards made it even easier. I barely looked at the screen and always looked at my audience.

The medical board went along as expected. I turned in all my papers from Emory showing all the things that were wrong with me and kept going to school. We were going to be gone so we started packing and getting ready to leave for Rock Hill, South Carolina, her hometown. I met her parents when they visited Phoenix sometime during the three years we stayed there, but never went to South Carolina until I became Med boarded. I could tell my marriage was falling apart. I thought if I deserve being treated like this there must be a good reason, but my memory returned and there wasn't. Maybe she got fed up with me and basically gave up on me. I was being treated badly by her like I was not part of the family. The kids no longer looked up to me and she had control of the money, so we were always broke somehow. It was very depressing. While in my office I checked my email and found one from American Ninja Warrior asking me to apply. I immediately start crying realizing how long this road to recovery is and how far I have to go physically and mentally. I emailed back, telling them I could not but I will apply next year. This gave me a sort of goal to keep in mind and it kind of planted a seed of persistence.

Chapter 17

Medical Retirement to Rock Hill, SC

My rating came to 100 percent. I was really messed up to say the least. My pain tolerance was very high. I guess my brain still had some connections to make. I still couldn't feel my legs fully but they worked. I took advantage of this strangeness and got a tattoo. This tattoo took six hours straight. It is a phoenix across my whole back. Essentially this represents me. I am without a doubt a phoenix with every intention of rising from the ashes. During the tattoo, this was all I thought about. I was completely silent and in total focus, imagining me overcoming all odds. Everything was going to be OK. I hoped and prayed. I believed in God, I just didn't have a religion.

I met with Dr. Barrow in full uniform and he was impressed with my progress. He said wow, you look good. Just six months ago you were dying on my table. "We really can't live till we die a little, can we?" (Deadpool 2). Heather found a house for us to rent in Rock Hill and we headed there with her dad driving the U-Haul truck in October of 2013. The house was amazing. It was a four bedroom, two bath house with a huge master bedroom we put our bedroom set, treadmill, T.V and gym equipment in. On top of that we had a walk-in closet like in Alabama, so we had plenty of room. We unpacked and settled in. I couldn't help with the heavy stuff. It was six months after my surgery, so we had friends help us.

There was a Steelers game coming up in Pittsburgh, so Heather bought tickets and gave me a Steelers jacket. We moved in, situated everyone's rooms, and enrolled the kids in school. We flew to Pittsburgh and saw the game. They won easily. We got in an argument over something stupid, and it seemed when things were going too well she hated it. I tried to find a balance. I just wanted this to work out. Who knows if I would find love again. I found a sort of balance but it was more of a compromise, mainly because she knew I would do anything to make her happy and I was doing all I could. I was walking on eggshells. I told a pastor how our relationship was and he told me it sounded like we were unequally yolked and gave me a cross. We went to the gym together, ran, and had great sex. I didn't understand why we had financial and marital problems. Heather was no longer loving on me. It went back to how it was before we went to Benning. Nothing happened unless I initiated it. I felt unattractive and came up with theories of why she didn't seem attracted to me.

The staples got removed but I had this scare, maybe I'm not good enough in bed or could it be because I'm Mexican and her family doesn't fully approve. Every Thanksgiving a honey ham would show up on our doorstep. Every year, God is good. No matter what, Christmas was always good. I made sure the kids got what they wanted. Makenzie started developing the horrible attitude she saw her mom had half the time and was very ungrateful. She would act like a fool as she watched the others open presents, she would yell and scream about their presents and would not even look grateful opening her own.

In January we went to New York and had a great time. I don't think we argued at all. We went to see the Kelly and Micheal show, visited Trump Tower and the Statue of Liberty all in one day. The rest of the time we enjoyed ourselves at nice pizza spots and strolled through Central Park watching performers, and had to visit a five star restaurant of course. We took pictures of us making angels in the leaves. I remember thinking we were going to be OK and I love her so much. I was wrong. I don't know if in her past relationship someone had to be the dominant one, but I rather be equals. She went back to being domineering and uncaring. I didn't even get woken to participate in taking the kids to school. When I started, the kids didn't treat me like a dad anymore. Often I spent time alone. We didn't watch shows and movies together anymore. I just went to the park and listened to Tony Robbins' Limitless Passion, Les Brown and Bruce Lee on YouTube to start.

Everybody seemed to hate Trump so I bought his book *Make America Great Again* to understand why. The only reason I could really see it is because he wasn't a career politician. He disturbed the usual cycle of a career politician, taking office and doing absolutely nothing for the country like Obama and Biden. It was strange to see how the media turned a blind eye to everything that went on. It seemed like they had selective attention and hearing, and I stopped trusting the media. Then, I stopped watching all together. I thought back on the Travon Martin and Casey Anthony incidents, all misleading either on purpose or accident. All their credibility was lost. It's like watching the pied piper play tune after tune, but I'm not dancing to it.

We continued living a fruitful life in Rock Hill, where we bought a four foot above ground pool and a nice grill that I cooked on while everyone relaxed in it. I love cooking for my family and they loved my cooking. We decided to have her tubes untied because I was hoping having a kid would bring us closer together. We had it done and she wouldn't get pregnant for some reason, but we kept trying. Everyone had a phone, we had cable, we ate out often, and we had all three kids in soccer. This was all on my income that she had control of, which is why we had little money and stopped traveling. Things were bad. I felt unloved and I couldn't believe the woman that I fell in love with seemed to hate my guts.

I thanked God and Jesus every single day, still do, as well as prayed for my marriage. I wrote my papers and had her proof read them before I turned them in. I had other methods of editing but I wanted her to be impressed and proud of me. I guess but I never got that impression. I got compliments from my teachers on my work, so I knew I was doing good work. I was satisfied my GPA was 4.0. I finally graduated in 2015 and had my ceremony in June, in which my mom Belinda Garcia brought my nana Virginia Lugo all the way from Phoenix. We made homemade tamales and watched movies. No one saw her domineering uncaring side but me and the kids.

We had plenty of fun and it was great for the time being. As soon as they left she was back to being hateful Heather all over again. She also gave off this negative vibe that made me uneasy cause I would be the target of negativity. I felt like I was walking on eggshells too often. This is not a way to live. "The poorest way to face life is to face it with a sneer" (Teddy Roosevelt 1838-1919). This is basically how she was in Phoenix until we left. Until the surgery, she didn't act like this again. But this did happen to us, not just me, I thought to myself. Heather did something that brought us closer together and signed us up to be GOTR (Girls On The Run) coaches after asking me. Besides the attitude maybe she wanted to save our relationship also.

This after school program had Belle and others in the same age group or grade that we trained to run and do activities. I found out that I'm really good at motivating them to run, even the most out of shape. I had them running well in no time. Heather would tell me "I don't know how you do it." To this day a few are on my Facebook. Activities like the beach ball, where they hit it as many times they can without touching the floor, is on my YouTube channel, Google #sapper_ninja. This ended and we started drifting apart again. I was applying for Ninja Warrior and working on my grip strength as well as my upper body strength. Just trying to focus on something other than my situation. I planned on working on my truck also. To one up me, Heather said she would add a T.V to hers and other things, so I settled for a new bike.

I always wanted a Red-line bike, so I got a royal blue one, my favorite color. Heather was pissed anytime I spent some money on myself. Living like this was horrible, but I got a call from somebody telling me to apply for Ninja Warriors military qualifier, and got my info to prove I was a veteran. I told Heather and she didn't believe me and started telling me I got scammed. Couldn't she just be happy for me? As time went on I started to believe her till one day I got the call and was asked to compete in the show's military qualifiers in California. I told Heather and started crying, so she asked why. I said no matter how good my story is, if I fail, they will never call me again.

To make matters worse I broke my fifth metatarsal in my right foot doing tricks on my bike and was in a cast, which I decided to cut off myself ahead of time to compete. She sent Belle with me to California. We got there a few days early and went to check me in. I was worried about my foot. That pinky toe bone was fragile but it was a price I was willing to pay. My brother Kaye, who gave me my nickname Sapper Ninja because it seemed I could do anything in the army, gave me a pep talk on the phone and I was good to go. The other ninjas were doing something called the flag pole and I thought it looked cool so I tried it. You hold the pole and extend your body like a flag. When I did this, my old boxing injury was not in mind, not until my left shoulder dislocated. It freaked out everybody and the medical staff popped it back in by telling me to hold a weight with my fingertips and it popped back in. This was some adverse shit. I can do it. This is what I thought to myself as I showed the crew I could still do this, climbing on stuff. See, I'm good to go, you don't have to worry, I told them.

Heather arrived with the other kids to watch the show. My foot was the least of my concerns now. It was my turn to run the course. I hit the quintuple steps perfectly until the last one. On the last one I held the step briefly and my left arm dislocated. The crowd gasps then yells as I make my way to the platform at the end. I was able to pull on the padding with my finger tips, popping my arm back in place as they screamed even louder, wildly even. I thought to myself, this was definitely going to be on T.V. I had to jump from a trampoline to a cargo net and didn't make it. I didn't want to swing my arms. Afterwards the cameras were in my face asking me how I think I did, and simply said I did good. All this never aired.

I returned to Rock Hill and back to the walking on eggshells life I had. I had surgery to fix my shoulder and I applied for 2016 as soon as it was open and made my own video. My other one was done by a pro. Christmas came around and everybody got what they wanted. Brandon loved his PS4. We had a family gathering at her moms and she started being hateful. On the way home she started saying mean things to me because I was talking about Christina Ricci's acting with the fam. The kids joined in, I lost it and screamed "I don't want to be in this family" and got out of the car to walk home. I walked around the park thinking, how do I get out of this? The dogs are gone. They went to the pound after eating a dog. All I had was Marley, my new dog. I raised her to be like a person, my daughter if you will. Nothing was keeping me here but my ability to drive away. I thought suicide was the answer so I went home and got my gun and walked right back to the park with Marley. I sat there with my 40 caliber XDM on the bench for an hour on the verge of shooting myself and Marley looked so worried. I decided not to go through with it and went back home.

I applied for Ninja Warrior for 2016 and opened my own bank account at Bank of America to build up money to do this ninja stuff. When my wife found out she got really pissed and told me she wanted me out of the house. I was on my computer trying to find out how to do this and an article popped up about Palmetto Harmony and the founder Janel Ralph. This company made CBD oils for epilepsy and it worked for her daughter, so I reached out to her on Facebook. I gave her my number and told her my situation. She called and asked for my card info. Within days I had it and was taking it, instantly noticing a difference. I soon passed my usual seizure time. I was having one to two every two months before and would chip a tooth sometimes or chew my tongue up. There are two types of seizures to me: the one's I see coming and the ones I don't. The ones you see coming are the scariest. You see flashing. It's like you're going to turn off and die and everything shuts off one at a time. The ones you don't, though, are very dangerous and less scary for me, but also scary to everyone else.

I reached out to pastor Mike Cook to train at his course, and he trained me to do the salmon ladder and other things. I told him what I was going through and he prayed with me. He suggested I make a post lifting her up and praising her on Facebook. I got a call from her yelling at me. After seeing my application video, pastor Mike told me I should join the walk on line in Georgia with John Brown, so I went. When I got there I was asked if I wanted to be the first in line and I said yes. I was so excited that I called Heather and told her and asked if she would come see me mid March. I gave her all the info and told her it would make me so happy.

The day came and I kept calling her to see if they made it and no answer. I was first in line waiting to run the course and they came to interview me. I was asked if I was scared. I said after Iraq, not much scares me. Then they say "who do you have in the stands here to watch you run?" I lost it and broke out in tears. This shocked them and they stopped the interview. I ran the course and failed of course. You need to be in a good mental state or you fail. I was heartbroken. I was just able to hide it until he asked that question. I guess I'm good at hiding my pain. Afterwards one of the ninja's asked me what was going on, so I told him the situation between me and my wife. He said he had a house he could rent me nearby and I told him if I can find a way, I will. I got back and had the will to fix my truck, get my license, and trained my dog to sit in the passenger side. She thought I was bluffing; she knew I feared being alone all over again. I pleaded to her one last time, saying how good we could have it if we got along, just loved each other, and be grateful.

Chapter 18

Going to Gresham, GA to Survive

My heartfelt plea was just laughed at and my heart sank to my stomach. I called my ninja friend and told him it was on. I will see him the next day to get the house and made calls to stop her caregiver payments. You see, she got 3500 a month to treat me like this and thought I couldn't stop it. A few calls fixed that and I closed the joint bank account. I went to Planet Fitness alone before leaving and this lady asked me about my head, so I told her about everything. Then she knew I was leaving due to my situation and told me she was a member of The Quilts of Valor Foundation. She arranged for me to receive one before I left. The one who presented it to me was Harvey Mayhill. I also attended a GOTR reward ceremony where we all got these pink coach capes. "Where there's a will, there is a way."

On April 20, 2016, I put what clothes I could in garbage bags and put my bike in the truck bed. My T.V and electronics were in the backseat. Marley sat in the passenger seat. That's all I took. More than half of my movie collection was left behind, but I took what I could. I made it in about four hours and I felt refreshed and relieved. I already survived another dark chapter in my life. I moved right in after finding somewhere to eat. He gave me a tour around the house and it was under renovation a bit so there was one bathroom. Kim told me about the ninja gyms in the area- Ninja Quest, Ninjakour and Slingshot Entertainment- so I tried them all and went to Ninja Quest most of the time. At this house, I was alone and sleeping by myself for the first time in 11 years. I could not sleep so I drank wine every night and still do except on Sunday. It's hard to fall asleep alone. The house was stocked with wine and Kim said help yourself. I stayed away from the white due to acidity.

In May, not long after I had left, I got a text from my mom asking who some guy on her post was. Apparently she deleted me and everyone I knew off Facebook but her. I made a new profile and found she was seeing this guy before I left. It seems like the sneering attitude meant she was ready to cheat on me or was already. I was so mad I jumped in my truck to go work out and crashed in the back of a semi. I was driving to and from until then. I got the truck fixed for more than I bought it but I needed it.

In June, the University of Phoenix contacted me and was so inspired by my accomplishment despite the odds. They were very surprised to see what I was doing now so they sent a crew to do some video of me training. They put many videos on YouTube of me training at Ninja Quest and walking on the slack line in my backyard. This can easily be seen if you goggle #sapper_ninja. I told Heather. I don't know why, I just wanted her to love me again. We started talking again but I wasn't sure what would come of it. Maybe you just don't know what you have till it's gone. I was going through all my movies and I started watching *Saving Private Ryan*. At the end, he says "earn this." I thought about Unruh and realized I don't need to feel survivor's guilt anymore. I've earned it. I am a good man doing good things in this world and you get out of this world what you put into it.

The Fourth of July was fun. I took my dog to the park for the show, she barks at them but she's not scared. I bought plane tickets to visit Phoenix for my birthday and planned on having Marley boarded while gone. Heather and I kept talking it out and she decided to come be with me in Georgia in August. She left that plumber guy and came to stay with me, and it was great. We started The Love Dare book from the movie Fireproof and I was so hopeful again I thought it was going to be OK. I was in the process of filing for my CRSC (Combat Related Service

Connected) pay. I was being helped by Cori Cabell, my Wounded Warrior representative, and we were happy. Heather wanted me to set up caregiver pay again and a joint account so I started. Soon after I got the CRSC pay in a lump sum. It was a significant amount so I paid her Dodge Journey off completely.

My birthday came up so I left to Phoenix and even came back early, I didn't want anything to go downhill. She started being hateful and domineering once again, and I couldn't believe it. Did she sense I ran into money and began acting like this? I tried to figure out why someone would throw away a good thing like this but I just kept trying. Rather than continue writing about my mistreatment, I will move on to the next chapter. Just keep in mind during the next I'm dealing with her and basically groveling for love and affection. Being ungrateful for what you have just gives you less in life to be grateful for. "I was too weak to give in, too strong to lose" (Foo Fighters The Best of You).

Chapter 19

The Ninja Community

The ninja community saved my life, and as soon as I got connected with Mike Cook I was introduced to them. First was at the walk on line in Atlanta, and his ninja competitions where I got baptized before leaving Atlanta, solidifying my relationship with God becoming a Christian. The ninja community was very friendly. They knew what I was going through and invited me to events like parties and competitions. I even had some come over from time to time. Some smoked weed and sold it to me. One even taught me how to slack-line, that is balancing on a long ratchet strap while walking. This was very hard at first. My ninja Nick brought a 100 foot one for him and a 30' one for me.

At first I couldn't even balance, but by the time Heather showed up I was walking forward and backwards easily. Many of the ninjas are focused on health, so I discovered so many health related things from vitamins and minerals, to oils and muscle stretching. Nick is a professional masseuse, so I paid for a deep tissue massage for sure. I needed it due to my muscles knotting up from overuse. I was strong but this was a whole different area of strength to get better at. Balance, swiftness, arm and grip strength seemed to be the key factors, so I worked on those. I drove to the ninja gyms often even though I have epilepsy and half my vision went to the left on both eyes. Imagine driving back at night. I went so much I eventually mastered every obstacle at the gyms, from the warped wall and spider climb to the floating doors.

I had a membership to both ninja gyms and went to slingshot often since they had the tallest warped wall. The hardest obstacle I mastered was the floating doors, which took a while to improve my grip strength enough to do it. The first obstacle I mastered is the warped wall since it is the biggest one. Once a week Ninja Quest had an obstacle course run that I had Heather take me to. Once she showed up in August, I asked her to drive me around, which she was not happy about, so I began to feel like a burden. I wished she saw me as a blessing. One time one of the ninjas drove me home after seeing me driving at night. It was dangerous. They would set up a new course and I would go as far as I could every time. These courses let me know how my progress in my ninja skills was, and grip strength has always been my weak point. Heather was not impressed, but it was OK with me. I was relieved that she took me.

I was training so much my muscles were knotting up and I was developing injuries. I didn't stretch my legs and hip flexor, so I pulled it running up the warped wall and my achilles tendon got very stiff. It is important to stretch every day regardless of what I'm going to do, so I start every day with a stretch since Atlanta. I also get a deep tissue massage from time to time since then to loosen me up. The first four months I learned so much from the ninja community as I trained and focused on myself and self care. I even started drinking different water. I prefer PH water now. I trained on obstacles like the salmon ladder and wind chimes as well. There are many others but I can't really remember since I don't watch that reality show anymore. Being at so many recordings in different locations gave me a different perspective of it. The physical feats are real, but how real is a reality show anyway, he says while winking.

I would get invited to celebrations by people in the community too, at restaurants and at the gyms also. It felt like I belonged there. This was a way different environment than what I was used to. I was with a family that I cared for, despite being fractured from combat, that didn't care much for me after I got broken. Now I was around people that seemed to care about my well being, enough to invite me to stuff since they knew my situation. During the whole eight months in Georgia, I became way stronger as I was preparing for the applications for the ninja show to open. They didn't show me in 2015 or 2016 when I walked on, but I still planned on applying and getting in the walk on line as well. I had plenty of money, so I showed up nice and early, always first in line on the list.

I was a walk once in Georgia and four times in California, but went to many shows regardless. I went to several shows all over the U.S and many Vegas finals also. I've seen enough to know better. I did like being a tester though, because there was no pressure. We just tested and they made adjustments and we did it again. It was so fun, especially testing in Vegas. Ninja Quest was my favorite ninja gym because they had the most obstacles. The owner, Glenn Davis, was always adding something new. I would train as much as possible on the new ones, making the most of my gym membership. Ryan Stratis also worked there, which I thought was pretty cool. He was on the show often at the time. Pastor Mike Cook had competitions I went to often, so I was training for those as well. I would hang out with the other ninja also and just walk around and go to movies and stuff like *Deadpool*. I can really relate to that movie. "Life is an endless series of train wrecks with only brief commercial-like breaks of happiness" is my favorite quote from the movie.

We also went to this spa called JeJu Spa and got massages, relaxed in the many different saunas, and got in the hot tub. It was a great place to recover. There were at least seven or more saunas and like five hot tubs. I had plenty of space at the house, so I let ninjas that came for competitions stay there free of charge from time to time. I went to every course night and every competition. It relieved the stress of the situation I started going through at home. I started setting up the caregiver pay but stopped when her attitude changed, as I wasn't ever doing a joint bank account again. It was the only time I was ever in the negatives. That made her mad. She said, I'm your wife. I shouldn't have to work. I should be on the account. I signed up for the ninja competition in January on November fifth, so I was preparing for it. Thanksgiving didn't go well. I had already got Christmas gifts. I told her this was not going to work out and got her a Uhaul truck and helped her move. They got their presents but I decided I did not want to spend another holiday unhappy. I wasn't going to let her break me. I was flowing like water already. "Empty your mind, be formless, shapeless, like water. Put water in a cup it becomes the cup, put water in a teapot it becomes the teapot, be water my friend" (Bruce Lee).

Chapter 20

Rebuilding My Life in Phoenix, AZ

I couldn't drive across the country by myself, so I flew in. We call my mom's handyman, Mike, "Mike on a bike." He helped me pack and connect the truck to the u-haul. I told Kim I had to break the lease in February and he was OK with it. By December 15th I was gone to spend time with my family in Phoenix for Christmas. Christmas was great. It was the first happy one in a long time. I wanted to get involved with something because I'm dedicated to doing good things with my life.

I got involved with an organization that gave out stuff to the homeless. The Joe Niekro Foundation had a group meeting for AVM survivors that I went to often. Once they heard my story, they wanted me to tell it at their next Phoenix Knuckleball: A Pitch for Life Gala 2017. I spoke at the event and told my story in 2017, and they were very moved. I began focusing on my health and self care, watching what I ate, building an obstacle course in my backyard, and taking care of my body. I started with my feet. Years of constant and consistent neglect gave me bad athlete's foot. I thought it was callused skin, but under it was bad athlete's foot. Pedicures helped but I started using a foot peel mask, which cured it after a few years. Constant and consistent effort, whether it be good or bad, is the reason for being overweight, underweight or overall good and bad condition. I applied for the ninja show coming up in 2017 and went to walk on line in Cali.

As I suspected, I didn't get a call but that didn't matter. I had the second or third in line on the paper we handed the staff. We would meet and form a list. After that I went back to the Banana Bungalow and checked in by text once a day on messenger. I also played around at the ninja gyms in the area just to stay ready. The day of the show was finally here and I was ready to run. I watched carefully as the others ran. My turn to run was here. Seventh walk on to run, I got through the quintuple steps one at a time to be safe since people seemed to be falling on them. Cannonball run was easy, and they kept chanting hold on tight and I breezed through. Once past the steps, and that ball next was the fly wheels with grips to hold on to. I jumped too hard and split my right eyebrow open, which poured blood down the side of my face. I didn't notice until I got to the last wheel and looked down to see blood dripping in the water. I smiled and laughed a bit as I continued my progress to the next obstacle.

Next was the block run. I blew through that without any problems and perfect balance that hyped me up. The battering ram obstacle was next until I got to it. Nobody had passed it yet. It was three free spinning cylinders on a horizontal pole: small, medium, and large. The last one I had a little trouble with, but I just shimmied along to the end. The warped wall was the last obstacle for qualifiers. This was my first time doing the warped wall on the show. I was careful not to touch the blood, keeping my grip solid until they threw me a rag to wipe my head. I ran up, grabbing the top with one slightly bloody hand. Before I could bring the other up, I slipped, and after two more tries I was done. I went far enough and fast enough to pass qualifiers and advanced to the semi finals.

Afterwards, I walked to my mom and Vince, my biological father, and hugged them on the sideline. Vince lived in California, so he showed up to see me run the course. They took me backstage and did some clips of me saying what they wanted me to say, but I wouldn't play along so that didn't last long. The next day I blew through all the obstacles even better and faster than before, killing the wall in one run up. I should have taken a rest for my arms because I went up the salmon ladder and fell. I went up three rungs and felt it, so I rested while they chanted one more rung. That last rung got me. They did not show that awesome run, just a small clip of me falling from the salmon ladder. The only time they aired my run and I had to bleed for it to be shown. Obviously they cared about success, not progress or my story, but I kept applying and getting in the walk on line.

I got my medical card which was good for edibles, but I got my weed by the pound elsewhere. I continued adding obstacles to my course and bought trussing to add even more to it. I was getting so much better at this ninja stuff. I took a real estate class to switch things up as I prepared for the next show in 2018. Trump was president and the country was doing great. He was severely hated but I liked him. He was doing a great job. Bubba and I had to go to Joe's Halloween party so we asked for a ride from my mom. We went as Mario and Luigi. I was Luigi and he was Mario. We had a blast. Joe has his own house and a family now. Only a few of the homies accomplished this. Most of them were dead or in prison.

My mom got to her business party late and grabbed a raffle ticket. They did the raffle and my mom won a cruise ship trip for two by Princess Cruise and she took me along. We were booked for a Crown Princess ship in June, which would sail to five places in Europe: Genoa, Rome, Italy, Pisa, Paris,

France, Gibraltar, and Barcelona, Spain were on the list to cruise to and we had our passports ready.

I applied to the ninja reality show for 2018 and got in the walk on line, ensuring my chance. I stepped up and ran the Jurassic park themed course like a champ. After the steps were the jumper cables, two red dowels, or nunchucks you grab to swing at a bag. I grabbed the bag with my legs, pulled it in and rode the bag to the end easily. They turned the block run to a ball run, so I ran on big red spheres instead of blocks. Next was the skyhooks, which were holding a ring and jumping from hook to hook using it. My grip failed me. Into the water I went on the last hook. Grip was always my weak point. That run was not shown either, but it was good. I started to notice people from better parts of town, like Scottsdale and Glendale, got calls often. It was the right side of the tracks, so to speak.

I got back home and continued training that grip by dead hanging as long as possible. June came and we went on our trip. First stop Rome. There was so much to see. The Colosseum and all the other architecture was amazing. We had club class on the ship, so we had a large room with a view, and when we ate our personal chef served us when we went to eat. Next was Pisam and on the way there one of the commentators of the show posted a pic of me doing the ball run. I showed my mom and assumed they would show me, but didn't. I wasn't bothered that my life was getting so much better.

We got to Pisa and saw so many sculptures and intricate buildings, not to mention the great food. The leaning tower of Pisa was scary to look at. Next was Genoa, Italy. My mom always made lasagna with hamburger meat, never sausage, and that's how they do it in Italy. My mom and I ate it at this nice restaurant outside, in front to enjoy the weather. Marseilles was a shortstop, but we saw what we could before we had to be back to the ship. Gibraltar was the next destination, it was Spain but controlled by the British. We saw the rock of Gibraltar and took the tour as well. The tour was the best part.

As we walked along, these monkeys called Macaques with no fear and much curiosity, came out and followed the group. When we stopped, they would come up very close. I gave one half my banana to one and ate my half next to him. One jumped on my mom's head and started searching her hair, so I took video of it and also took pics. I was laughing so hard, telling her to stay still and don't freak out. Barcelona was amazing. We watched the magic fountain Montjuic for quite awhile before walking around and checking out the many cathedrals and flamingo dancers.

Chapter 21

Repairs on My Mom's House

Something I forgot to mention is, as soon as I arrived, I started repairs on my mom's house. The first thing I had repaired was the sliding glass door. My brother was pushed into and almost through it. My mom boarded it up and it stayed like that for years. I had Window World come and replace two windows at a time. That one and the kitchen window were first. Next was the bathroom. I had a new tub shower head and toilet put in. Even a new toilet, as the ones we had were from the 50's. After the bathrooms, I decided we needed a better fridge, so I bought a new top-of-the-line fridge for her. We picked out $350 of groceries on the Sam's Club app and filled it.

Next, two bedroom windows were replaced. They did it quickly, in like 30 minutes tops. I love Samsung products, so I bought her a Samsung washer and matching dryer. To finish off the windows, I had three done: two bedroom windows and the living room, all double-paned. This allowed the house to retain its temperature way better, costing my mom less money. I was buying the groceries, so she saved even more money. This led her to the decision to remodel the kitchen, from the cabinets to the floors even. New cabinets, stove, oven,dishwasher, and floors, which now looks like wood, but its tile. Last but not least, I asked my mom how old the air conditioner was and she said about 20 years old. It was time to get rid of it and replace it altogether with no repairs. I did my research and decided to get the very best one I could find. I got a Trane air conditioner. None of this was financed, I just bought it.

Chapter 22

Growing Stronger in Phoenix

That trip was an amazing blessing from God. It was that good, believe me. Two years back I was in a much different situation, but now I was good to go for sure. I was 33 or so by then, and have been through and experienced so much. I feel truly blessed looking back at the way I took on every challenge and obstacle in my life.

In 2019, I decided to switch things up once again and took a stock market class. This proved to be a good move since people were about to be led into a COVID-19 pandemonium by the media. Whoever decided to invest in that pied piper nonsense made good money on Moderna while watching the masses freak out. A recurring craziness seems to happen nine to ten months before the election, and it's ridiculous, but this time I watched the stock market. All I knew is there was no way I was voting for Biden after seeing Obama's run.

I thought I could find love once again since I officially divorced, but I just slept with many women. I want to connect but it seems like women are more interested in one night stands. In California, I woke up in this beautiful woman's bed for the first one night stand while there for the show. I busted my head in 2017 and thought it was cool. After a while it was just meaningless sex. In Phoenix, I had many more and just two relationships that are not worth talking about. One is crazy and one lies to people. Tricia is still a good friend of mine that comes over from time to time. We go out to eat and walk around and talk while Marley runs about at the park. We also have great sex, "Dripping wet as we experiment in sweaty sex" (Tupac 1996).

One girl started coming over and we would talk and smoke weed often, so I thought she liked me. My mom said she just wants my weed. To see if that was true, I offered her some for oral sex and she said no. So she really liked me, I thought, but when I started smoking just a joint with her she stopped visiting. I have to be cautious with women. It seems most just want to take advantage of me in some way.

What I want for you to learn from this book is that whatever point you are in life, you are always the author of your story, so get out there and experience life by setting goals and learning all you can. Staying persistent and consistent will get you there, just keep going. "The world is behind us once a motherfucker get an understanding of the game, and what levels and rules of the game is. The world is no trick no more, it's a game to be played. So now we lookin at the world from behind us. Ninjas know what they gotta do, just have to put your mind to it and do it, real ninja's do what they want to do bitch ninjas do what they do" (Tupac, "Starin At The World Through My Rearview," 1997).

I plan on buying a house close to my mom's, but until then I am good to go. I offered to buy the house next door, but he knew I had money so he tried to haggle with me. I started training the kids in the neighborhood on my course and they were getting really good. All I required was that they maintain a C+ or better in school. This gave me something to do. Everyday I had them come at five pm when the sun wasn't scorching the course. I tested their balance, grip strength, just like they did on the warped wall and wrote it down. Each kid had their own PT card to record progress. This went on until the covid crap. The kids asked me if I wiped down the course and I laughed and said no. That was the last time they trained them. Out of all emotions, fear is perhaps the most dangerous.

One day I found out Drew Drechel was arrested for sleeping with an underage girl. At a time when all the media was putting everything they could out, none put that out there at all. It was very strange, and as I thought about it, I was very puzzled. I decided to step away from that ninja stuff. I didn't want or need that fleeting moment of success. The ninja community that sprouted due to it helped me, so it served its purpose. There were so many awesome ninjas that didn't get aired. I now live at my mom's and I do not watch T.V. programming, because it seems that programming is society seat fillers, running around, smashing things. These people had to because they were bused in. Craziness. At one point it had people wearing masks everywhere, but not me unless I flew. I got tired of it and stopped flying.

I got baptized a second time in my hometown of Phoenix, AZ by pastor Mike Williams. I hope to go to Phi Phi island when this idiocracy that started in 2020, that strangely looks like the plot for V for Vendetta, passes. I never let them stick me, enough is enough. It is the island in that movie *The Beach* with Leonardo Dicaprio. My mom is a bit of a tyrant, so I'm looking and waiting for a nice house to be available in the area so I can make it even nicer, or a fixer upper. My friend from elementary school, Andrew Tressler, knew I was a GOTR coach. He has a son in baseball named Bryce. He wanted me to train. So I started training him to sprint, starting with a video of Usain Bolts' running technique. Then I recorded his sprint time as a base. He learned the slack line as well. Every day at five he sprinted and I critiqued him until baseball started back up and he made junior varsity.

Every November, I take out $200 and put $100 on two peoples groceries. I like helping people and doing good. I take my dog Marley to the park and enjoy my life. I spend time learning new things and cooking all kinds of different recipes for my mom as well as Bubba. One day I will have a family of my own again. I have never had a credit card but have always had plenty of money since I was 16. Systemic racism is an excuse for the lazy and self-entitled. This is not the end of my story; there is more to come. This is where I am so far. Who knows what the future holds, but God has a plan for me. I give $100 a year to the Wounded Warrior Program in Unruh's name, so I can put the donation card or coin when I have time and the means to visit.

I don't spend my time watching the market too much, but if I see an opportunity, I take it. Otherwise, I feel I'm one of those people that can't peel themselves away from the slot machine. "Our deepest fear is not that we are inadequate. Our deepest fear is that we are powerful beyond measure. It is our not our darkness that most frightens us. We ask ourselves, who am I to be brilliant, gorgeous, talented, fabulous? Actually who are you not to be? You are a child of God, your playing small does not serve the world. There is nothing enlightened about shrinking so that other people won't feel insecure around you. We are all meant to shine as children do. We were all born to manifest the glory of God that is within us; it is in everyone. And as we let our light shine we unconsciously give other people to do the same. As we are liberated from our own fear, our presence automatically liberates others" (Marianne Williamson). I first heard this while watching the movie *Coach Carter*. Fear will enslave you, but knowledge is power, power you can use to fight that fear. "You cannot oppress the people who are not afraid anymore" (Cesar Chavez).

I am rebuilding my movie collection one movie at a time since I have rebuilt my life here. Lately, I got American Me, all three Hangover movies, my favorite John Singleton movieHigher Learning, Free Guy, Chappie and American Sniper. It is about Chris Kyle, the Devil of Ramadi, as the enemy called him, I have read. I also bought Goodfella's, American Gangster, Carlito's Way, Casino, Get Rich or Die Trying, and Scar-face of course for my gangster collection. My collection of religious movies are God's Not Dead one and two, A Light In The Darkness, God's Not Dead: We The People, and Passion of The Christ.

It seems like I may need eye surgery on that right eye once again, as it has double vision now. I noticed it while training Bryce Tressler to sprint. Besides the drinking, I take very good care of myself with vitamins, minerals and oils. I just keep researching and finding out so much about these. I now have a full sleeve of tattoos on my left arm. From shoulder to wrist is the Engineer Castle with a Sapper tab above it, the American flag around my bicep with the Great Pyramid in the empty space, and a UFO above it. Below that is Goku from Dragon Ball Z, and next that is the Eye of Ra and Second Infantry Division patch. On my forearm is the rose that grew through the concrete, the symbol for Jeet Kune Do, and Si Vis Pacem Para Bellum. If you want peace, prepare for war. "If you believe, then you can achieve: just look at me" (Tupac "Smile," 1997). "Without struggle there is no progress" (Fredrick Douglas). When you focus on getting and having, you won't become much of anything. When you focus on being and becoming, you will have so much. That is all I have to say about my story so far.

AUTHOR BIO

Daniel Moreno, a strong Mexican American born in Phoenix, AZ on August 22nd, 1983, has gone through many struggles but has overcome them and used them to grow and become stronger.

Made in the USA
Monee, IL
08 November 2026

67382866R00125